ALSO BY A. PETER KLIMLEY

Great White Sharks: The Biology of Carcharodon carcharias
(ed. with David G. Ainley)

A. PETER KLIMLEY, PH.D.

THE SECRET LIFE OF SHARKS

A LEADING MARINE BIOLOGIST REVEALS
THE MYSTERIES OF SHARK BEHAVIOR

SIMON & SCHUSTER
NEW YORK   LONDON   TORONTO   SYDNEY   SINGAPORE

SIMON & SCHUSTER
Rockefeller Center
1230 Avenue of the Americas
New York, NY 10020

597.315
KLIM

SIMON & SCHUSTER and colophon are registered trademarks of Simon & Schuster, Inc.

For information regarding special discounts for bulk purchases,
please contact Simon & Schuster Special Sales at
1-800-456-6798 or business@simonandschuster.com

32530 60564 4002

Manufactured in the United States of America

10 9 8 7 6 5 4 3 2 1

Library of Congress Cataloging-in-Publication Data
Klimley, A. Peter
    The secret life of sharks : a leading marine biologist reveals the mysteries of shark
    behavior / A. Peter Klimley
        p. cm.
    Includes biographical references and index.
    1. Sharks—Behavior. I. Title.
QL638.9.K57 2003
597.3'15—dc21   2003042726
ISBN 0-7432-4170-3

ACKNOWLEDGMENTS

I have written many technical articles reporting new observations or the results of experiments for my scientific colleagues during my 30-year tenure as a marine biologist. These articles are by design concise, objective, and devoid of emotion. Their content is limited to a brief description of the methods used in a study, its results, and their relationship to the historical body of scientific knowledge. The articles always seem to me to lack the story behind the facts. For that reason, I have always wanted to write a book about the adventure of being a scientist and seeking new knowledge. I graduated from college with a passionate ambition to explore the world. After a brief stint traveling to remote places on land and finding them already populated, I decided that the oceans were still a frontier and being a marine scientist was a great way to explore. What excites me most about being a scientist is how you go about finding out a fact, not the knowledge itself. One is challenged to meet new people, travel to distant places, build innovative devices, put oneself in harms way, and develop novel ideas. I have written this book to give you, who are not scientists by profession, an idea of what fun and excitement there is in scientific exploration. This includes not only the joy of observing something that no one else has seen and explaining it in an entirely new way, but also experiencing the pain and generating the conviction needed to defend your ideas when they are either misunderstood or disbelieved. This book chronicles an exciting quest to study sharks and two particularly charismatic individuals—the hammerhead and white sharks. The book may catch your fancy for that reason. However, you should also know that every scientist experiences the same passion and shares the same excitement when investigating a natural or physical phe-nomenon, be it as simple as the rainbow produced by light passing through a prism or as grand as the creation of the solar system.

I would like to thank Jim Wade and Bob Bender for the editorial guidance they gave to me when writing this book. I acted on the bulk of their advice, and the book was much improved. One of the first persons to whom I confided an interest in writing a book about my experiences studying sharks in their underwater world was Richard Milner. He is a successful author who had written a popular biography about Charles Darwin. I got to know Richard when he edited my arti-cle in *Natural History* magazine about the social behavior and naviga-tional ability of hammerhead sharks. He recommended that I contact his book agent, Ed Knappman of New England Publishing, and ask him whether he would be willing to help me locate someone willing to publish such a personal account. He contacted Bob Bender of Simon & Schuster, who liked the chapters I had already written and agreed to publish the book.

I want to thank my wife, Pat Klimley, for first reading the chapters and suggesting improvements before submitting them to the demand-ing eye of a professional editor. I would also like to express my grati-tude to the many students, scientists, and professors who have participated in my many expeditions and cruises to learn about shark behavior. Several organizations—the National Geographic Society, the National Park Service, the Office of Naval Research, the National Sci-ence Foundation, and SeaWorld—have provided funding for my sci-entific studies of sharks. Finally, three of my teachers—Art Myrberg, Don Nelson, and Dick Rosenblatt—were particularly influential in my growth as a marine scientist.

I would like to dedicate this book to the memory of my father and mother, Stanley and Dorothy Klimley, who were wonderful and caring parents. It is our parents who are our first and foremost teachers. They make our adult accomplishments possible with their affection, example, and knowledge. The scientific odyssey described in this book would have never happened without their support.

CONTENTS

CHAPTER 1

SHARK FEVER

It was 1985—a decade after *Jaws* set box-office records across America—that I first saw a stunning predatory attack by a great white shark off the coast of California. It was early in the morning on a gray, cloudy October day. I was checking the motor on my boat, which sat on a concrete platform leading to the edge of a cliff at Southeast Farallon Island, 30 miles from San Francisco. In an hour, the boat would be lifted off the platform and lowered onto the water 25 feet below, and then I would commence my search for white sharks in the waters surrounding the island. But my first encounter came a great deal sooner than I expected!

The only sound that morning came from the waves breaking against the jagged rocks of the island as I stood there planning my day's schedule. Suddenly, there was an almost explosive burst of activity among the gulls, which began flying in a circle close to shore. Then I noticed that the gray water 25 yards away had changed to a vivid crimson, stained by the blood of a struggling northern elephant seal. A huge shark, much of its body out of the water in clear view, swam vigorously back and forth in the pool of bloodstained water. Its animated movements made it seem as though it was enjoying its meal. This horrifying sight sent a chill down my spine and brought a cold sweat to my forehead. Although all of us regularly have the feelings of thirst, hunger, anger, and sexual arousal, we rarely experience the fear of being eaten by a predator. But I felt this heart-stopping emotion when watching that white shark devour a seal.

Since witnessing that first attack, I have analyzed more than a hundred video records of white shark attacks on seals and sea lions. In 1987, my colleagues and I established a continuous watch for these natural feeding events off the South Farallon Islands. Volunteer biologists, stationed at the peak of Lighthouse Hill, scanned the surface waters around the islands during daylight hours each fall for an explosive splash or bloodstained water at the surface that would alert them to a shark attack. The attacks that volunteers witnessed from 1988 to 1999 were recorded on videotape. I have listened to the voices of these people on the tapes trembling as they describe a shark feeding on a seal. The narrated videotapes are shocking—full of blood, guts, violent struggle, and death in the open water. And yet these accounts, combined with the results of other research on the white shark, reveal that our conception of the white shark's feeding behavior, based largely on *Jaws*, is more fancy than fact. The white shark is not a dumb feeding machine, but a skilled and stealthy predator that feeds with both ritual and purpose—and may prefer not to eat human flesh.

Let me briefly describe a white shark predatory attack on a seal. Both the reports of observers and their videos of attacks show a pattern markedly different from popular myths about shark behavior. None of the observers witnessed or recorded the most frightening feature of a shark attack, that powerful, bone-crushing first bite. This suggests that the shark seizes the seal below the surface. Typically, an observer first spotted gulls circling over a large bloodstained area of the water. The bloodstain often elongated in one direction, and then the shark appeared with the seal beside it. Or the shark often swam with unusually wide and sweeping tail beats. Such beating would be necessary for the shark to propel itself forward if it was carrying a heavy seal in its jaws. After a prolonged interval the seal usually rose to the surface and floated there motionless with a sizable chunk of flesh torn out of it, although the wound was no longer bleeding. This last observation was puzzling because seals contain large amounts of blood, used to transport oxygen to their tissues. This plentiful supply of oxygenated blood enables a seal to dive as deep as 1,600 feet and stay down over twenty minutes.

The shark then surfaced quickly, swam to the carcass, and seized it. This scenario led me to believe that the white shark kills its prey by exsanguination—it allows the seal to bleed to death. The shark had to hold the seal tightly in its jaw until it was no longer bleeding, and only then would it bite into it and begin feeding.

The videotapes also provided answers to other puzzling questions—matters of more than academic interest to those who want to survive a shark attack. How does a shark decide what and when it is going to attack, and how does it determine what it is going to eat? While observing sharks feeding on seals in the waters at the Farallon Islands, only once was there an attack on a human being. The attack occurred at 2 P.M. on September 9, 1989. It began like many of the attacks on seals. The white shark seized the leg of Mark Tisserand, a commercial abalone diver, while he was maintaining a prone position and trying to clear his ears 25 feet below the surface, a couple of hundred yards from shore. According to Mark, the shark swam up from underneath, seized him, carried him, pulling him down for six seconds, and then suddenly let go and swam off. He said that he bled profusely while in the jaws of the shark. This pattern appeared to me consistent with the notion that the shark exsanguinates its prey, but what struck me as odd was that the shark released its human prey intact and swam off. A shark would have chased down a seal and promptly finished it off.

The fact that Mark hit the shark with the butt of a bang stick, a short metal pole with an explosive cartridge at one end, might have caused the massive shark to let go. But the pattern he reported, in which he was bitten and then released, is typical of white shark encounters with other human beings, usually those who were not lucky enough to have a weapon in hand. Could white sharks release people because they find them unpalatable? In another attack at the island, a shark seized and released a brown pelican, even though the bird was quickly disabled and unable to resist further attack. This tactic was consistent with another perplexing observation: many sea otters are found along the Santa Cruz coast each year with fragments of white shark teeth embedded in their open wounds. Sharks often lose their teeth when biting down on prey, and so they have many rows of teeth in their jaws that can move forward and replace those lost. A sea otter has yet to be found in the stomach of a white shark. Why didn't the sharks eat these tasty morsels?

It struck me that there was a connection between these observations. Birds, sea otters, and people are composed mainly of muscle, whereas the preferred prey of the white sharks—seals, sea lions, and whales—are composed mainly of fat. Could the sharks prefer energy-rich marine mammals to other comparatively energy-deficient species? During the fall of 1985, I had repeatedly tried but failed to interest sharks in consuming sheep carcasses, which have little fat on them. These sheep carcasses had electronic beacons concealed within them that could be used to track the movements of the sharks, once the beacons had been swallowed. On the other hand, white sharks *will* swallow seal carcasses with beacons concealed within them. And, significantly, in one case a shark refused to eat the muscle of a seal with its fat removed. I became more confident of this preference for fat after viewing videotape of a shark scavenging on a dead whale floating at the surface. The shark opened its large mouth, propelled itself forward so that its upper jaw reached far up on the body of the whale, closed the upper jaw, and ripped a large amount of fat from the body. It left the muscle of the whale.

Why should a white shark be a picky feeder, and why show a preference for fatty prey? The answer may be that surplus energy is needed to keep the shark's body warm. I remember once placing my hand on the back of a large white shark as it moved like a locomotive alongside of my boat. Its body was warm, around 75°F in contrast to the cold water, often in the 50° range off Northern California. In the body of the shark are retes, or clusters of veins and arteries, which act as a heat-exchange mechanism to warm the brain and musculature, increasing the transfer of oxygen to the tissues and quickening the reaction times of the nervous system in the cold waters the sharks inhabit. Perhaps high-energy food contributes to the maintenance of the shark's warm body temperature. It probably helps the shark grow fast. Adult white sharks increase their body size more than 5 percent a year. This growth rate is twice that of the porbeagle shark, a fish-eater in the same family that also inhabits cold, temperate waters, and three times that of the mako, another fish-eater from the same family that lives in warm, tropical waters. The white shark is actually a maverick among sharks, quite different in its behavior and physiology from the many tropical species. It has colonized the relatively shark-free cold, temperate waters where seals are abundant because by gorging on the layers of fatty seal tissue, the white shark derives thermal energy from the same insulating fat that enables seals to survive and thrive in cold waters.

I have devoted much of my adult life to finding out what the white shark and other species of sharks do and why they do it. I have discovered that sharks have a rich and complex repertoire of behaviors. Yet most people think they are simple and reflexive. This belief has arisen out of ignorance because people have avoided being in the water with them for fear of being attacked. Unlike most humans, I have sought to enter the water and observe sharks, not to rush out of the water onto a beach like the terrified swimmers in *Jaws*. In real life, the more you know about sharks, the more you respect them for what they really are. Yes, like any other animal, I have a core of fear of sharks in my more primitive animal mind, a fear that probably is as old as human consciousness. And I have a healthy respect for what sharks can do when I am on their turf, so to speak. But I swim with them and study them because I find them utterly fascinating.

Certain things that happened in the early years of my life made me increasingly interested in understanding the behavior of the creatures that inhabit the oceans—but I had no idea in those years that I would spend most of my life studying such formidable creatures.

I was born on March 7, 1947, in the city of White Plains, New York, 25 miles north of New York City. I spent most of my childhood in Lakeridge, a White Plains suburb. When I was very young, my family vacationed during the summer in a cottage on the sandy shores of Peconic Bay on eastern Long Island. During these formative years of my life, I developed a keen interest in marine life. I walked along the banks of the salt marshes at the edge of the bay with a net in hand looking for blue crabs and shuffled my feet along the shallows feeling for the hard shells of cherrystone clams buried in the sand.

When only eight years old, during spring 1955, I swam with mask and snorkle in the clear waters of the Mediterranean. I speared my first octopus and persuaded the cook at the hotel to cook it for me. This diving experience furthered my interest in the ocean.

As an adolescent in the 1960s when my friends were going to the beach and listening to surfing music, I was busily raising and watching tropical fish. I spent as much time staring at my half-dozen fish tanks in my bedroom as some of my friends did staring at their television sets. Here was this microcosmic society of fish exhibiting behavior just as diverse and complex as the behavior of family and friends.

I found the mating ritual of the Siamese fighting fish, native to the swamps of Southeast Asia, particularly mesmerizing. The male would bend his body and long, flowing fins around the female, squeeze to force eggs out through her vent, and release a cloud of sperm to fertilize them. As the eggs slowly floated toward the bottom, he would swim downward, gather the eggs in his mouth and rush to a nest of mucus-coated bubbles that he made at the surface of the water. He then would blow the eggs out of his mouth into the mucus nest. The eggs would hatch a few days later into tiny baby fish. Their bodies were clear with conspicuous yellow sacs of yolk on their bellies that provided essential nutrition during the first days of their existence. Remaining below the nest, the male would catch newly born fish that fell out of the nest and return them to their proper place. What an exemplary parent!

Another favorite of mine was the male African mouth-brooding cichlid. The male in these species inhales the fertilized eggs of the female and keeps them in its mouth for several days. Opening and closing his gill slits, he forces water past the eggs in order to ensure the ample supply of oxygen needed for the eggs to develop. Once they hatch, the young use the male's mouth as a refuge. Twenty to thirty may come out searching for food and spread out within the tank, but when frightened, they immediately form a tight school and swim into Daddy's mouth for protection.

Males of both these species are highly aggressive. Indeed, Siamese fighting fish earned their name because it was popular in Siam to wager on the outcome of fierce battles between two males placed in the same fishbowl. Initially, the males position themselves side to side and unravel and extend their long flowing fins, appearing twice their normal size, to intimidate their adversary into retreat. Often they beat their tails toward the other, driving water against the other fish's sensitive body. Confined to a small container, the subordinate is unable to hide and the displaying is replaced by actual combat. The males ram each other, dislodging scales and tearing fins. African cichlids match them for aggression. They patrol a small area on the bottom of a lake and chase any intruders from their territories. These fish usually do not bite an intruder as do the Siamese fighting fish confined to a small fishbowl, but they do put on the same aggressive display, just as we might clench and raise our fists to show we are ready to fight. The male cichlid also turns its side toward another male and spreads its fins to appear large and formidable. Fish, I decided, were not dumb creatures after all, but displayed the rudimentary equivalents of many complex human behaviors. Peering into that aquarium hour after hour, I got hooked on fish. It was high time for me to learn how to function in their world.

When I turned thirteen in 1960, I joined the local swimming team, which practiced and competed in the muddy waters of our neighborhood lake. I won the fourteen-and-under butterfly race in the Westchester County Championship in my very first year of competition, and I continued to swim competitively in preparatory school and college.

In my junior and senior years of college at the State University of New York at Stony Brook, I began to concentrate on my science courses. At this time, I worked as a volunteer for Bill Rowland, a graduate student in the Biology Department (now professor of animal behavior at Indiana University), who was studying the behavior of stickleback fishes. I helped him collect fish and perform his experiments, and we became close friends. He rekindled my passion for observing the behavior of marine life.

A year and a half out of college, in spring 1972, I married Patricia McIntyre, whom I had first met in a swimming clinic when I was thirteen years old and she was fourteen. She had won the backstroke race in the Westchester County Championship the year before, and spent the practices swimming back and forth in the pool doing the backstroke. I had won the same competition in the butterfly, and later backstroke, and divided my time between both strokes during practice. I began to date her when I was sixteen. She was my first boss, head lifeguard at a local swimming pool. We dated off and on until we married. She has always been a strong supporter of my consuming ambition to study sharks, and it is unlikely that I would have succeeded in this crazy dream of mine without her constant encouragement.

She has participated as a volunteer in many of my studies (as she has a career of her own), and has learned not to fear sharks. When she first joined me in the water to observe schooling hammerheads, one of the cruise members remarked to me that she swam so close to me that her long snorkel tube seemed to intertwine around mine. By coincidence, we experienced some peril that day when a huge blue marlin, over 10 feet long, swam up to us and directed a threat display toward us. The marlin turned to the side, raising its dorsal (top) fin and depressing its pectoral (side) fins to reveal its full size, and rapidly opened and shut its mouth in an ominous fashion. We backed slowly away in order to reduce its stress (confronted with a choice between fight or flight), and prevent it from dashing toward us and skewering us on its long, sharp bill. There is some chance of pushing away an onrushing shark before it bites you, but little chance of deflecting a charging marlin with a swordlike 4-foot-long bill. Billfish vigorously defend a personal space around themselves, and I was well aware at the time that a stout bill had been removed from the hull of a submersible that ventured too close to a marlin.

I spent the 1972–73 school year teaching marine biology to teenagers aboard a square-rigged ship that sailed throughout the Caribbean. My first encounter with a live shark occurred while spear fishing for groupers, a common fish that inhabits the local coral reefs. I was swimming at the edge of a vertical submarine cliff dropping 1,000 feet into deep water off the island of Grand Bahama. I had speared two groupers and was swimming back to shore, dragging the fish behind me attached to a line around my waist. Suddenly, a slender gray shark with a slightly yellow hue—a lemon shark—less than 5 feet long, started circling me. I marveled at how it swam, sinusoidal waves slowly passing down its body and tail in a fashion reminiscent of the way a snake moves on land. But the shark always approached me from the rear, where I couldn't see it easily and push it off with my hands, which made me a little uneasy. The shark was attracted either to the body fluids of the fish coming from the spear wound or the tiny water displacements created as I moved my feet. It eventually dashed toward the fish on the line behind me. Alarmed by this, I moved my arms back and forth while blowing air out my mouth in order to distinguish myself from ordinary "prey" such as the fish on the end of the string. This frightened the shark momentarily. However, it still followed me all the way to shore.

The lemon shark had approached a little too close for comfort, but it hadn't tried to bite me as sharks' sinister reputation had led me to fear. It just wanted a free lunch by stealing the fish I had speared. Thinking about it later, I realized this behavior made sense given that the shark's mouth was less than 8 inches across, wide enough to swallow a fish but not big enough to manage to grip and rip an arm or leg off my body.

My second shark encounter occurred later that year while I was spear fishing off the coast of Panama. There was a dispute between the owners of the square-rigger and the school, and the ship was abandoned in the Panama Canal while litigation ensued between the two parties and teaching was suspended temporarily before continuing ashore. I joined the first mate and two other members of the crew and embarked on a voyage to the city of Cartagena, Colombia, on a small whaling boat. The first mate had converted this boat into a Viking ship by adding a mast with sails, gold-colored wooden shields along the sides, a dragon tail on the stern, and a dragon head on the bow.

We were running low on food by the time we passed the San Blas Islands off the southern coast of Panama. I volunteered to spear some fish for dinner. I lowered myself into the warm, clear water wearing a mask, snorkel, and fins and began searching for edible fish on the reefs below. It took less than ten minutes to spear two large groupers. I threaded the line attached to my weight belt through the mouth and gills of the two groupers, tied the end of the line to the base of the line, and began swimming back to the boat. Suddenly, two massive gray sharks approached; they probably weighed twice as much as I did and they were longer than I was by a good two feet. The sharks' bodies were broad from the head to the base of the tail, so they swam more stiffly than the sinuous lemon shark. Both sharks had huge, blunt heads—hence the species is commonly called the "bull" shark. The sharks were hungry too, and were interested in the groupers on the line trailing from my belt. They continued to swim directly toward the fish on my stringer. This time I let go of the line. One of the monsters slowly swam up to a grouper on the end of the line, opened its large mouth, and swallowed the 20-pound fish whole. The shark's partner quickly ate the second grouper on the line. The pair then lost their interest in me. Even these large sharks did not appear to be the menace that I had expected.

Before my first two shark encounters in the wild, my perception of sharks, like that of most people, had been shaped by nature film documentaries that almost invariably showed sharks in pools of blood, their gaping mouths filled with many rows of sharp teeth. The impression the films left was that sharks were dumb, instinctive feeding machines and humans were their favorite snack. But the sharks I encountered in the Caribbean didn't seem to be man-eaters. I began to wonder if these beautiful creatures could exhibit complex behaviors just like the tropical fish back home in my aquariums. Did sharks have complex mating rituals? Did they fight to the death or settle disputes nonviolently with aggressive displays? I guessed they must have some form of ritual combat because otherwise both participants in any encounter would be mauled, given their formidable armament of sharp teeth.

I was twenty-three, and less than two years out of college, when I started on my pursuit of sharks while teaching marine biology to high school students aboard the square-rigger mentioned earlier. It was berthed at the Albury Docks on an island crossed by a freeway leading to South Beach in Miami. At this time, I was 6 feet 1 inch tall, with long brown curly hair and a pirate's mustache. During the day, I wore blue jeans, a multicolored tie-dyed tee shirt, and a denim jacket, and in the evening, a long black cape. In other words, I was a typical hippie.

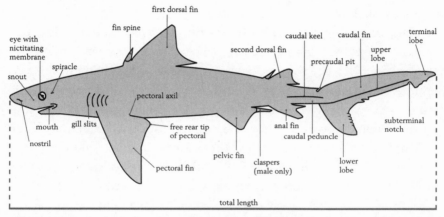

The anatomy of a shark. (REDRAWN FROM CASTRO, J.I. 1983. *The Sharks of North American Waters.* TEXAS A & M UNIVERSITY PRESS, COLLEGE STATION)

One fall day, I took my class to the University of Miami's Rosenstiel School of Marine and Atmospheric Science. The school is a cluster of buildings sitting on a small coral island, halfway between the coast of Miami and Key Biscayne. We marched into a building full of plastic pipes and large tanks for keeping marine animals and plants alive. Sonny Gruber, a recent graduate, showed us a little lemon shark peering out of a small wooden tank with a Plexiglas bubble at one end. We crowded around the tank. A blue-green beam of light was trained on the eye of this small shark and a mild shock was applied to its body by two electrodes inside the tank. It blinked its eyelid, but to the students' astonishment the eyelid rose upward and covered the eye from below and not from above as in humans. Furthermore, the eye appeared to glow when the light was shined on it. Sonny explained that this was because there were mirrorlike granules in the back of the eye that reflected light back through the retina with its light-sensing cells. This improved the sensitivity of the eye and enabled lemon sharks to see really well at night. Eventually, the shark would learn to blink in response to just the beam of light; the shock wouldn't be needed to elicit the response. Sharks could also learn! Sonny found that the sharks reacted differentially to colors. The small lemon sharks were most sensitive to light of the blue-green wavelength. This was the wavelength that penetrated the ocean depths farther than such others as red or orange.

His face lighting up with excitement, Sonny said he was now trying to determine whether sharks saw colors as humans might see them. The U.S. Navy was underwriting his research. To Sonny, this was an interesting scientific pursuit, but it was hardly an abstract academic issue to the Navy, which worried that sharks might be attracted to life-jackets colored yellow to make them more visible to airmen in rescue helicopters. Sailors jokingly called the color "yum-yum yellow."

I asked Sonny to have lunch with us on the sailboat, but he couldn't because of a conflicting engagement. He suggested that as a replacement I invite his colleague, Arthur Myrberg, an expert on the hearing capacities of sharks. Art gave us a wonderful lecture. He had just completed a study focusing on the behavior of a small hammer-head shark, the bonnethead, in a shallow tidal pool exhibit at the Miami Seaquarium, a large aquarium next to the marine laboratory.

Art and I instantly hit it off. We were both keenly interested in animal behavior and sharks. Art had been a prize pupil of the great Konrad Lorenz, one of the first ethologists. (In Greek, the word "ethos" means "character" and the suffix "-ology" means "knowledge." Thus scientists who study the character and behavior of animals in their own environment are called ethologists.)

Art had just completed an ethogram for the bonnethead, a species of shark possessing a laterally elongated head with a rounded edge resembling a lady's bonnet. An ethogram in this case was a catalogue of swimming movements and postures for the species—the first step to understanding its social order. Art urged me to undertake graduate studies and learn from him the classical techniques used to describe animal behavior. Before I could even think about it, he was already outlining a research project for me—to describe the behavioral reper-toire of one member of the species known as requiem sharks, so termed because they are known to attack humans. The blacknose, a small, slender gray shark with a black patch on the end of its snout, is one of thirty or so larger species of requiem sharks and a member of the scientific genus *Carcharhinus*. These species mainly live in tropical waters, often around reefs, and are also called reef sharks. This species was an ideal subject for a behavioral study because it was small enough to live in the confines of a shallow tidal pool exhibit at the Miami Seaquarium. So, with encouragement from this eminent scientist, I entered graduate school at the University of Miami in 1973 to begin work on my master's degree.

I was never able to establish a colony of blacknoses in the tidal pool exhibit at the Miami Seaquarium. I spent much time fishing for them in their favored habitat, the shallow flats off Miami and the Florida Keys, with no success. I captured only one shark over a period of six months in the waters off Grassy Key, halfway down the Florida Keys between Miami and Key West. This shark died while being moved from the Keys to Miami despite my transporting it in a large circular rubber tank, perched on top of a flatbed truck. The water in the tank had been collected from the habitat of the shark, circulated with a bilge pump, and infused with oxygen to help the shark breathe during its trip to Miami. After a Herculean effort, and much frustration, I gave up my dream of keeping this species in captivity to study its behavior.

Art—brilliant, intense, and incredibly hardworking—had another Navy-sponsored project going during the early 1970s; its objective was to learn more about the low-frequency, prey-generated sounds that attracted sharks. Eventually, my role would be to investigate the reverse—to examine the nature of the fright response some sharks show to sudden loud sounds.

For extra income, I worked two days each week as a research aide at *Sea Frontiers*, a magazine about the animals in the sea, verifying scientific facts in the articles submitted to the magazine for publication. Part of this job was to identify the species shown in the photographs that accompanied the articles. It took me no time at all to identify the white sharks in four photographs sent to the magazine to accompany a story written by Ron and Valerie Taylor, two members of a film crew that had made a sensational film about sharks, *Blue Water, White Death*, two years earlier. This movie depicted the adventures of a film crew that searched the world for the species called the "man-eater."

They finally found white sharks in the waters off Dangerous Reef, a small, flat island inhabited by sea lions several miles off the coast of South Australia. The first photo depicted a huge shark swimming in bloodstained water with its head and dark menacing eyes starting to come out of the water. The makers of this documentary first had to attract the sharks to the boat in order to photograph them. They accomplished this by pouring into the water a mixture of horse blood and macerated fish that Australians call "burley" and Americans term "chum." The two parts were mixed together in a barrel of water and siphoned into the sea day and night to create a "chum line," or chemical corridor, winding several miles away from the boat. With their keen sense of smell, white sharks quickly picked up this chemical trail—easy enough for them since the part of the white shark's brain devoted to smell is proportionately larger than that in any other shark species.

The second photograph showed a gaping mouth of a white shark, with its lower and upper jaws jammed between the heavy-duty metal bars of a protective cage in an apparent attempt to reach the person cowering inside the cage. Only after examining the picture closely did I recognize a piece of meat that had been attached to a metal bar in order to induce the shark to bite the cage. Clearly, this had been done to create the false impression that the shark was biting the cage in an attempt to seize the person within it.

The third photo was the most frightening of all because the shark, with its eyes opened wide, held a large chunk of meat in its menacing triangular teeth that jutted out of bloodred gums. The meat was attached to a rope, which was retrieved in order to induce the shark to swim beside the boat, where it could be photographed in a close-up view. This photograph was similar to one used to advertise *Blue Water, White Death*. On the poster, the gaping mouth of the white shark was accompanied by lettering in huge type that said, MAN-EATER. When you look at these photos, you imagine that you, instead of the meat, might be in the mouth of the shark—and that, of course, is exactly what the photographers intended.

During this same decade, the public's image of white sharks was even further distorted by Peter Benchley's book *Jaws* and the subsequent movie version, which drove home as relentlessly as the theme music the image of the great white shark as a dumb feeding machine. In the film, the shark's size was greatly exaggerated. It was almost as large as a fishing trawler, 30 to 40 feet long. The shark was also portrayed as an indiscriminate and voluminous feeder, biting the boat and trying to swallow 55-gallon drums. And it was given a vindictive nature, which compelled it ceaselessly to pursue the boat, its captain, and the marine biologist on board.

These two movies created a horrific image of the white shark that most people accept to this day. But these movies had a very different impact on me. From what I knew of fish and my limited experience with sharks, I just didn't believe this portrayal of the white shark. In fact, it struck me as almost a malicious libel and angered me.

It is unfortunate that we are often shown the white shark only during the brief, stunning moment when it is eating. Just imagine if a cinematographer wanted to make a documentary about your family banquet on Thanksgiving Day. He would remain in a corner of your dining room and record what you were doing using a video camera with a zoom lens. Every time a member of your family took a leg of turkey and raised it toward his or her mouth, he would zoom in on the item and the family member's mouth. It would suddenly open wide, revealing pearly white teeth in upper and lower jaws that would vigorously seize the leg and rip off a piece of tissue. The jaws would quickly move back and forth, chewing the meat, and the neck would widen, then narrow as the meat passed down the throat. As the camera panned around the table, focusing on each person's mouth, many humans would be shown feeding in the same way. Little other behavior would occur at the dinner table other than feeding. There would be little overt fighting among family members and little flirtation. This documentary about human beings eating at the dinner table might also be entitled *Jaws*.

In part because of my dawning awareness of the complexity of their behavior, I became more and more interested in learning about the real nature of sharks. I didn't want to study them in the man-made, artificial habitat of aquariums, however; I wanted to swim with them in their own environment, the ocean, where I might discover their true nature. I wanted to emulate Dian Fossey, who had gotten close to gorillas in order to study them. I had read in an issue of *National Geographic* magazine how dominant male gorillas allowed Fossey to approach members of their families unharmed because she imitated the submissive postures and movements of subordinate gorillas. There had to be some way to get close enough to sharks to observe them in their own environment without being attacked. Because I was a good swimmer and interested in fish, I wanted to study the most fearsome and powerful fish in the oceans. As my graduate education progressed, I began thinking about some of the questions I wanted to answer in my career as a marine biologist. Are white sharks the insatiable and indiscriminate feeders suggested by film documentaries showing multiple sharks repeatedly biting a cage? Do sharks really seek out humans as frequently as depicted in *Jaws*? I decided to see for myself how sharks behaved in the wild instead of as "actors" in a film.

CHAPTER 2

CROSS-SPECIES DRESSING

I would spend a significant amount of my graduate education at the University of Miami in a shark cage. I was the observer for a research team, funded by the U.S. Navy, with the task of learning as much as possible about why large sharks were attracted to pulses of low-frequency sounds.

During my three years at the University of Miami, I spent four to six hours a day, two days each week, dressed in wet suit, fins, and mask in a shark cage floating at the surface and looking for sharks in the Gulf Stream off Miami. I kept my lower body inside the door that opened from the top of the cage and bent at the waist so that my upper torso extended out over the top. From this jackknife position, I could keep my head in the water and look into the crystal-clear blue waters below. I usually held the side of the cage with my hands and slowly moved my body around to periodically view a 360-degree panorama below me. There were times when two or three hours would go by, and all I would see were one or two masses of sargassum algae despite the sounds emitted by a speaker below me playing continuously during this period. Occasionally, I'd reach out and grasp a clump in my hand and bring it close to me to see what was inside. I was always surprised at the myriad species that lived in this microenvironment in the open ocean. If you looked hard, you could see a small brown shrimp whose coloration exactly matched that of the brown mottled seaweed. You could also often find, camouflaged as were all of these Lilliputian species, a sargassum fish. With its dispro-portionately wide jaw, it could swallow a shrimp whole.

Forty feet below me was a big cylindrical, shiny metallic underwa-ter speaker, sending out low-frequency sound, attached to a cable leading up to the A-frame on the stern of the R/V *Orca*. This was a 40-foot research vessel (hence the initials "R" and "V") of the Univer-sity of Miami, used for daylong expeditions in local waters. The A-frame was a structure composed of two metal girders hinged to either side of the stern and welded together at the top to hold a pulley, through which passed the metal cable and electrical cord leading to the speaker. At the base of the girders were hydraulic supports that, when activated, extended outward, pushing the top of A-frame out-ward and over the stern. A nylon line from the shark cage was attached to a cleat on the stern of the boat so that I was floating on the surface no more than 60 feet away from the speaker. At the time, the research boat seemed much too far away and the rope much too thin. Aboard the boat were the skipper, our electronics technician, and a massive amount of equipment used to reproduce and amplify very low pitched sounds. When I leaned over the edge of the cage and put my head under the water, I could hear a monotonous, staccato drum-ming sound coming from the speaker: *borum-bodum-bodum-borum-bodum.* . . .

One day during the spring of 1975, I was watching six to eight slender gray silky sharks swimming around the speaker, slowly moving their long tail fins back and forth sinuously while gliding forward using their horizontal pectoral fins. Occasionally, one or two of these grace-ful animals would stop circling the speaker, accelerate upward toward me, and slowly start circling my cage, which was more like a large birdcage than those stout cages made of thick aluminum or stainless steel bars one sees in documentary films about white sharks. My fore-arms could extend over the edges of the cylindrical cage, which was 3 feet wide and 8 feet long and made by welding together wire, thinner than the power cable to your television, into a patchwork of squares about a hand-width in size. Despite the daintiness of my cramped cage and the fearsome nature of these sharks, I always felt quite secure, if sometimes a bit seasick. The sharks appeared more beautiful than terrible, moving gracefully below me with bright undulating light from the waves above reflecting off their backs.

My primary task was to count the number of sharks attracted to sounds of different bands of frequencies, some so low-pitched that they were barely audible to the human ear despite being broadcast from the speaker at powerful intensities. These shark-attracting sounds had been discovered by two of my University of Miami colleagues, Don Nelson and Sonny Gruber, during their research in 1963.

When he had been a graduate student at the Rosenstiel marine laboratory, Don had made extra money selling fish that he caught using a spear gun in the Florida Keys and the Bahamas. One day, he noticed that sharks usually appeared quickly when a fish was struggling at the end of his spear but were rarely seen at other times. Don hypothesized that the spasmodic movements of the dying fish generated a punctuated, staccato sound that not only consisted of pressure oscillations detectable by the inner ear of the shark but also miniscule jerking movements of water detected by the shark's lateral line. This sensory organ is visible on most fish; it consists of a canal leading from head to tail on either side of the body with the many small openings, which give it the appearance of a line. These tiny holes permit water to enter and exert pressure on tiny fingerlike receptors inside.

Don and Sonny, his best friend and fellow graduate student, made a tape recording of drumming sounds similar to those made by a struggling fish. They used a switch to key on and off a stream of white noise and passed this sound through an acoustic filter to eliminate the high frequencies. White noise is similar to the roar of the ocean and is composed of the frequencies audible to humans, high and low. Sonny then alternated between playing back this pulsed low-frequency sound and playing control sounds, which were not pulsed and at higher frequencies, while Don watched from a cage to see when sharks were attracted to the speaker. The sharks appeared only when the drumming sounds were played; they were nowhere to be seen when a variety of unrelated sounds was played. Later, Don and Sonny's degree supervisor circled above in his airplane while they played the sounds. He observed sharks, visible in the shallow waters, abruptly turning around and heading toward the speaker with great haste from distances close to half a mile away.

While floating in my cage, it occurred to me we should try a new experiment. There were so many sharks around the speaker that day that I felt we had surely collected enough data on how to attract sharks. Rarely did we attract more than three silkies at a time. Just a few days before, I had lost my usual alertness after spending four hours in the cage looking constantly for sharks and seeing none. My gaze had drifted downward toward the speaker. Suddenly, by chance I saw a shark propel itself upward like a rocket from directly below the speaker, bite the speaker, and accelerate downward out of my sight below the speaker. This all happened within a few seconds, and a shiver ran through my body as I realized that the attack directed at the speaker could have occurred while I was looking elsewhere. I might be bitten by a shark before I could react and defend myself unless I remained constantly vigilant on the job.

I wondered what would happen if instead of trying to attract sharks we tried to repel them by playing the shriek made by killer whales when collectively feeding on marine mammals. For someone who had been a graduate student for barely more than a year, it might seem a bit presumptuous for me to design an experiment. Art Myrberg, the project leader and my master's degree supervisor, had recently had a heart attack and was slowly recuperating in the hospital, leaving me temporarily under my own guidance. Hopefully, he wouldn't be offended if I took the liberty of playing back these piercing sounds to sharks even if he was absent and unable to oversee the experiment.

We had acquired a tape recording of this sound, which resembled a loud human scream, from a local whale biologist, who had broadcast it to gray whales as they migrated past San Diego in a large pod. He broadcast these sounds from his boat and observed how the whales reacted. The scream was three seconds long, and started with a constant tone of medium pitch, a *wheee*, followed by a brief tone rising in pitch, an *oooh*, and ending with an eardrum-piercing, high-pitched *weeeeh*. As soon as the sound commenced, the whales lifted their heads up out of the water and looked around for their feared predator, the killer whale, and for a safe refuge. The whales then rapidly swam as a group to the nearest forest of kelp, an underwater plant that forms dense forests of stems and leaves rising to the surface in shallow water. The whales remained there motionless until the sounds ceased. The biologists also broadcast other sounds, similar but not identical to the killer-whale scream. Apparently, the gray whales were unafraid of these sounds because they did not swim away and hide. One sound that had little effect was a tone of a constant single frequency, termed a "pure" tone, of the same narrow frequency and energy content of the first *wheee*. Another sound was white noise of the same frequencies as the scream. Significantly, a mirror image of the sound, starting high, gliding downward, and ending at a low pitch did not frighten the whales. Their inattention to this very similar sound indicated that the whales had in their minds a clear acoustic image of the sound that scared them. The whales were not simply startled by a loud and sudden sound, as a person might be if a car were to suddenly screech to a stop nearby.

Two days after my brainstorm, Charlie Gordon, our electronic technician, and I set up the equipment and I floated off in my shark cage in the usual way. Our plan was to alternate playing the scream with the pulsed attractive sound. I waited until six of the sharks were swimming near the speaker and signaled to Charlie to play the scream. To my utter amazement, the sharks around the speaker abruptly turned and accelerated away like rockets. I shouted to Charlie, "It's incredible, they've gone!" Not only were sharks near the speaker frightened, but also two silkies swimming close to me near the surface turned and fled with such force that they produced huge swirls of water at the surface that impressed Charlie, who shouted, "Wow!"

I returned from this trip with my mind full of ideas. If the sounds of an adult killer whale, an occasional predator of sharks, were frightening, perhaps the sight of a little killer whale, who might (as far as the shark knew) be accompanied by a shark-eating mom, would also scare the sharks.

During my animal behavior class I had studied the work of two ethologists, Konrad Lorenz and Niko Tinbergen, both of whom had won the Nobel Prize for their studies of animal behavior. Ethologists are keenly aware of the importance to animals of a "key stimulus"—an innate, or genetically ingrained, image contained within the mind of an animal at birth that will trigger specific actions when an animal confronts the object. Lorenz and Tinbergen developed the theory of the key stimulus, which I thought might explain why newly born sharks were attracted to the pulsed sounds of their prey, even though they had never heard nor seen them before. A student at the nearby Mote Marine Laboratory had shown, as part of his research for his doctoral thesis, that baby lemon sharks, which had yet been in contact with struggling fish, were attracted by similar sounds.

According to these two pioneering ethologists, these images were not always perfectly formed in the mind at birth. For example, a male red-breasted robin defending his territory might mistake a red mop for the red breast of another territorial male. A killer whale is certainly an easily recognizable creature given its shiny black body, huge white H-shaped patch conspicuous on its belly, and two large oval white patches on either side of its body near the head. Furthermore, these fearsome, predatory whales have a large erect fin on their backs, jutting upward with a straight-edged front and a sicklelike rear edge. There are small dolphins in the genus *Lissodelphis* that look almost identical to the killer whale with the same white H and oval markings. I wondered whether these species of dolphins might have evolved their color pattern because the disguise might confer immunity to attacks by sharks fearful of killer whales. The shark might mistake the dolphin for a juvenile killer whale and be uncertain whether Mom was around and whether she was hungry for a meal of shark. I headed to the library and looked through scientific literature on killer whales. There were a few recorded incidents of killer whales eating sharks, but it was not a common occurrence. The whales spend most of their time in temperate and polar waters inhabited by their preferred prey, seals and sea lions, rather than in the tropics, where most sharks live.

To test my theory, I decided to play the role of the killer whale in my first scientific experiment. This was less time-consuming and certainly much cheaper than building a model of a killer whale. The real thing was out of the question, since all the killer whales that might be used in my experiment were busy performing tricks throughout the day in shows at the Miami Seaquarium and SeaWorld.

I did know of the whereabouts of some adult lemon sharks on which I could test my theory. They were being kept in a shallow fenced-off enclosure in the Florida Keys by Gerry Klay, who collected them and maintained them in seminatural confinement before sending them off to sea parks around the world. Here was the ideal site to conduct my experiment. But first I must make my killer-whale costume.

Upon returning to my home, a large sailboat, I pulled out a beautiful new wet suit and hood I had recently bought, placed it on the dock, and outlined a large H on the stomach and ovals on the neck and head. My wife and I lived at this time on a 36-foot wooden sailboat that we were in the process of renovating. White paint left over from painting the sailboat was on hand as was some quarter-inch plywood piled on the dock for future repairs of the deck of my boat. I cut out a square piece to support the fin. With a picture of a killer whale in one hand, I used the other to carefully draw the shape of a fin on the remaining plywood. After cutting out the fin with a jigsaw, I stood the fin up on the flat plate and joined the two with screws. The next step was to make two holes on the top of the backpack and two on the bottom, through which a couple of black ropes could be passed and joined by knots in order to hold the fin on my back. The final step was painting the fin a glossy black to resemble the dorsal, or top, fin, of the killer whale.

That afternoon, I went over to the Miami Seaquarium and watched how the killer whales swam in their aquarium. The success of my experiment would rest on closely imitating their movements: porpoising up and down, rolling from side to side, and beating their huge horizontal tail flukes up and down to propel themselves forward. I decided to use my long black skin-diving fins to imitate the tail flukes of the whale, holding both fins together and moving my legs up and down in the "dolphin kick" used by competitive swimmers. Having been a varsity swimmer and captain of my team in college, I found this part easy and fun.

When I practiced swimming porpoise-style along the beach in

front of the marine laboratory, the preoccupied scientists walking along the pier headed toward their research boats did a double take, first thinking they might have seen a bottlenose dolphin feeding near the beach and then realizing that it was only Art Myrberg's new graduate student. They must have wondered what kind of lunatic Art had brought into their institution.

Once I had mastered the necessary techniques, I headed south to Gerry Klay's place on Grassy Key, about halfway between Miami and Key West. Gerry was a tall, massive, and fearless Dutchman, who had become a shark collector only after working as a mercenary in New Guinea. Gerry loved to tell stories of his dangerous escapades, such as when he joined a band of headhunters in a war party. I could just imagine Gerry, a different skin color and twice the height of everyone else, shaking a spear in his hand as he stood in the line of warriors facing their adversaries. He could have frightened the enemy because of his size alone. He loved danger, and sharks fulfilled his need for it. They also provided him with a healthy income. We both loved sharks and happily talked about them for hours on end.

Gerry was obviously surprised when I arrived at his "Sharkquarium" with my killer-whale costume under my arm. He later admitted that my getup had struck him as a bit whimsical and that he had been astonished the United States Navy would actually pay a graduate student to do something as wacky and risky as this.

On the day of the experiment, Gerry was very eager to record the outcome of my meeting with three large and nasty lemon sharks that he now kept in the lagoon. He ran into the house to pick up his camera so that he could photograph me during the experiment. He surely did not want to miss getting a photo of me being attacked by one of his sharks! He could sell it to the *Miami Herald* and would have made the front page that Sunday.

I got into the wet suit on the concrete platform next to the lagoon. Putting on a wet suit always takes a bit of time and effort, but one with a killer-whale fin takes even longer. Gerry took my photograph as I stood in front of him, my fingers making the V symbolic of peace, as if to request that the three sharks behave nonaggressively. I grabbed a shark billy, a small two-foot-long broom handle with a nail driven into the end of the handle and ground down to a sharp tip. I would poke the nail into the snout of a shark if it tried to bite me. I then walked across the coralline rocks, sat at the edge while putting on my fins, pushed off, and began to swim toward the sharks at the far end of the lagoon.

Each time that I came to the surface after a shallow dive, I made a sound like *hiss-shoo* as if expelling air from my blowhole. This is what would be expected from a killer whale. All three sharks were resting on the bottom. The smallest one suddenly accelerated off the bottom, swam twice in a tight circle not more than 6 feet in diameter, and moved frantically alongside me while rocking back and forth and moving its jaws up and down in a menacing way. The second, slightly larger shark quickly followed the example of the first. However, it was the largest shark that, after circling twice, began to move back and forth in front of me in a looping pattern. His side fins were down, his back hunched, and his jaws were held wide open, revealing many rows of white teeth.

I understood this threat behavior well from previous encounters with other sharks (blacktips and bonnetheads) in a concrete tank in Gerry's backyard, and slowly backed away, using my hands in a sculling motion until I reached the safety of the rocks, on which I could climb to get out of the water. The aggressive behavior of the sharks was puzzling, but upon reflection, I realized that the sharks might be attempting to defend themselves because their ability to escape was thwarted by the small confines of the enclosure.

This was the kind of aggressive display that my colleague Don Nelson and his graduate student, Richard Johnson, had observed in the gray reef shark in the western Pacific island of Eniwetok. The lemon shark was trying to communicate a message to me, and this was "Keep away!" It contorted its body into a posture that made it look larger and opened and closed its mouth, showing me the weaponry it had in its biological arsenal. This behavior was similar to what I had observed when a neighbor's massive tabby cat cornered a small gray-haired cat of mine on the stairs leading to our house. My cat suddenly swelled to twice his normal size by erecting his hair and arching his back. He also opened his mouth to expose his teeth, hissed, and jutted his sharp claws out of his soft paws. The other cat's response was quick: it turned around and ran away as fast as possible with my cat in close pursuit. My cat stopped when he reached the edge of the yard and marched triumphantly back to me. How similar was the "display" of the lemon shark to the behavior of my cat on our doorstep?

What we needed now was a "control" to the experiment. The "variable," my killer-whale costume, had to be changed, but all other conditions had to be kept the same. I decided that the proper control was to take my killer-whale costume off and enter the lagoon with nothing on but my bathing suit. When I came close to the same three sharks, which were again resting on the bottom in the same spot, they ignored me and remained motionless on the bottom.

"Aha!" I thought, and donned my killer-whale costume again to demonstrate its obvious effect a second time. This time, only the largest shark was lying on the bottom on the far corner of the lagoon. When I approached, the shark again accelerated into tight circles and started coming at me, moving back and forth furiously in looping fashion until it unexpectedly dashed forward with lightning speed and tried to bite me on the head. I shoved my shark billy between the two of us and pushed him to the side. He tried to grab me a second time, however, so I wheeled around to keep the shark billy between the two of us. Eventually, I reached the rocks at the edge of the lagoon unscathed, but a shiver passed down my spine as I realized I had barely escaped being bitten by this shark.

When I got up on the rocks, I hastily explained to Gerry that my experimental design was all wrong. The problem here was that the three sharks were confined in the pen and had nowhere to go. Thus, when I intruded into their space, they were unable to flee. This was the classic situation of fight or flight, in which animals, including humans, if prevented from getting away from a dangerous intruder, will stop and fight furiously. It was obvious to me that I had forced the large shark into behaving the way he did because he was unable to flee outside the confines of the lagoon. What I needed to do was test the killer-whale costume in the open ocean where sharks would be unencumbered and free to escape from the killer whale.

My opportunity soon arose when Art—recovered from his heart

attack—and I were playing back the killer-whale scream to sharks off the island of Andros in the Bahamas. In a body of water called the Tongue of the Ocean, there was a large metallic buoy, 10 feet in diameter, which was moored near the island in water a half mile deep. Not only were silky sharks always near this buoy, but also pelagic whitetips, easily recognized by their rounded white-colored side and top fins. These sharks most often appeared in pairs, a male and a female. They swam so close to each other that at a distance their white patches merged into a single bright object. At the same time, the rest of their bodies blended into the background. This was because their bodies were countershaded; their backs, which were lighted from above, were darker than their undersides, which were shaded. This made the shark, when viewed from the side, a uniform blue-gray that matched the color of the surrounding water. All that was visible was a bright white object of uncertain identity. This type of object tends to attract fish in the featureless open ocean.

To Art and me, it appeared that the white patches on these pairs of whitetips together were serving as the perfect lure to attract prey to them. When we approached these lazily swimming sharks, they would suddenly accelerate forward with the most unexpected rapidity. We could just imagine how small fishes seeking refuge in the featureless ocean might be surprised and easily consumed by these sharks, who had deceived their prey by effectively mimicking a distant inanimate object.

We had come to the Tongue of the Ocean many times before to determine how attractive the pulsed sounds were to the sharks at the buoy. We would tie off the boat to the buoy with a long line and then lower our heavy underwater speaker below the boat. We would then summon the sharks away from the buoy to the speaker. On this particular day, I slipped off the end of the boat into the water and then quickly swam over to the shark cage. This was always a moment of anxiety, because I never knew whether a shark might be near. The water was crystal-clear, and not a shark was in sight. Attached to the cage was a Styrofoam buoy holding a small microphone and speaker, which I could use in communicating with Art and Charlie.

Our plan for the trip was to determine just how effective the killer-whale sound was at repelling silky and whitetip sharks. For that reason, a bleeding fish was attached to the speaker below me so that the shark would not only be attracted to sounds of prey, but also to the scent of its body fluids. Soon, I could hear the familiar sounds— *borum-bodum-bodum-borum-bodum*—emitting from the speaker.

Suddenly, three silky sharks were circling the underwater speaker closer and closer to feed on the attached fish. As the first came right up to the bleeding fish, I said to Charlie through the microphone, "Play the sound." Out of the speaker came the piercing shriek, *whee-oh-weeee*, of the killer-whale sound. All three sharks turned abruptly and dashed out of sight as if shot from a cannon. We had to play back the drumming sounds quite a while before they returned.

This time, Art wanted to increase their feeding motivation even more before playing the repellent sound. He instructed the crew to feed the sharks with our leftover lunch meat. I did not like this idea because the two whitetips became bolder toward me the more they were fed. The larger member of the pair kept trying to stick his snout through the 5-inch-square mesh and bite me, while I backed into the opposite corner of the cage. I looked at the shark, trying to get its snout inside the cage with its toothy mouth wide open. I reached over and tried to push its snout off the cage with my hand. The whitetip sharks acted so differently after being fed!

We played the scream as soon as the shark swam close to the underwater speaker, and this shark also dashed off. However, as we kept playing the scream, the sharks appeared less and less afraid of the sound. Our subjects were becoming "habituated," or used to the scream. This lessening of the frightening effect of the sound could have resulted because a killer whale never arrived after the sound was heard to provide the needed negative reinforcement, or because the shark anticipated the sound and was less surprised.

After we had completed testing sounds for the day, I tried out the killer-whale costume with silky and whitetip sharks. I let myself into the water with less apprehension, wearing my "sharkproof" suit with its white killer-whale markings and prominent wooden dorsal fin. Coming to the surface after each shallow dive, I did my blowhole imitation: *"hiss . . . shoo."* I carried the same short shark billy that I used with lemon sharks in the Sharkquarium, just in case a shark came too close. However, unlike the lemons, these oceanic sharks were swimming freely in the open water, not confined to a fenced enclosure. They were free to escape and, given this option, I thought, less likely to fight, or perform an aggressive display toward me, instead of fleeing. Would either the silkies or whitetips be repelled by the appearance of a killer whale?

My dive buddy, Don Wren, who made films for the popular television program *The American Sportsman*, swam somewhat hesitantly behind me carrying his large underwater camera. These bulky objects, held between them and the shark, always gave those camera guys courage. If my costume worked, it would be national news before the end of the week.

The group of three silkies kept circling me at a distance of 10 body lengths. The two whitetips actually were curious about the suit and came close enough to me that I had to push them off with the shark billy. I climbed out of the water and sat there wondering why the sharks didn't respond to the scream or my killer-whale getup. Why did they not have an adverse reaction to the sight of a killer whale when they were afraid of the species' sounds? Could it be my diminutive size? Although my fin and coloration resembled those of an adult killer whale, I was much smaller than an adult—weighing less than 200 pounds versus the 6-ton weight of a full-grown killer whale. Another possible explanation for the inaction of the sharks is that the withdrawal response was not instinctual after all, and these subtropical sharks had little opportunity to encounter and hear the hunting sounds of the killer whale, which spends more time in temperate and polar waters. There were other sharks in temperate waters such as the seal predator, the great white shark, that would have more contact with the killer whale and might have learned to avoid them.

Although I couldn't prove my theory about sharks and killer whales back then, I've been revisiting the concept and pondering a new experiment in light of new information that killer whales eat white sharks. During the fall of 1999, a killer whale was observed eating a smallish white shark, 13 feet long, at Southeast Farallon Island.

This rare and exciting meeting between these two species at the top of the food chain was observed by tourists on a day-trip to the islands to see the sea life. According to their eyewitness accounts, the whale carried the shark in its mouth at the surface for several minutes. I wondered: could the whale purposefully be killing the shark by holding it out of the water so that it could not breathe in its normal medium?

If so, it would certainly be ironic. As I suggested earlier, a white shark seizes a seal and appears to swim lazily with it underwater for five minutes or more while the seal loses a massive amount of blood. The seal is a master at holding its breath, and can stay underwater for over twenty minutes before coming to the surface. Part of this ability is due to the binding of air (or oxygen) with air-binding molecules (hemoglobin) in its blood. The shark's tactic quickens the death of the seal by depriving it of this air supply.

For more than a month after this attack, Peter Pyle, the head biologist during the fall season on Southeast Farallon Island, observed no feeding by white sharks on seals near the island. This was truly odd because for several weeks prior to the attack, sharks had been observed feeding on seals almost every day. Peter suggested that the sharks left the island because the killer whales were present. They may have done so out of fear of being eaten or because seeing the whales would cause the seals, the prey of the sharks, not to go into the water. It would be an exciting experiment to play back the screams of the killer whale to white sharks feeding on a seal to see if they would flee from the sound. In fact, it might be worthwhile putting on the killer-whale costume one more time and entering the water next to the seal and white sharks to see if the suit serves as a visual repellent. Of course, now that I'm a university professor I might want to be accompanied by two or more of my graduate students, also wearing killer-whale suits and carrying shark billies, just in case the costume doesn't work with great white sharks.

CHAPTER 3

SHARK SEX IN THE MIAMI SEAQUARIUM

Our research team initially believed that silky sharks in the Gulf Stream were frightened of the killer-whale scream in an apparently similar manner to gray whales migrating along the coast of Southern California when they heard this sound. We were uncertain whether the sharks responded only to the killer-whale scream and not to other loud sounds in the marine environment as well. Art Myrberg suggested that I conduct experiments aimed at determining what properties of the sound frightened the sharks. The response of sharks to the scream could be compared to their reactions to the control sounds used in the whale study. It would be easier to conduct these experiments on captive sharks in the Shark Channel located at the Miami Seaquarium than on sharks that lived in the open ocean, because a boat would not be needed and there would be less uncertainty involved in finding sharks to cooperate in my experiments. In this enclosure I had a captive audience of subjects! I devoted the next year of my life to playing back sounds to lemon sharks in the Shark Channel in order to answer this question. These studies led to my master's thesis at the University of Miami.

The Shark Channel was a large doughnut-shaped channel, 5 feet deep and 20 feet wide, containing two species of sharks that constantly swam around it. Within the channel at this time were a dozen lemon and three dozen brownish nurse sharks. The lemon sharks would be good experimental subjects because they were frightened by the coloration of my killer-whale suit in Gerry Klay's shark pen—I had yet to conduct the less successful tests of the suit on silkies and whitetips in the Bahamas. I would conduct these tests during the third and last year of my master's degree studies. The nurse shark was a common species in the waters off South Florida. Although sharklike in body shape, the nurse sharks have unusually long tails and large top dorsal and side pectoral fins. Protruding from the forward edge of both nostrils, which are located next to the mouth on the underside of the shark, are barbels—long, fleshy protuberances that may be sensitive to smells or touch and may help in finding food. The species name originated from the sharks' habit of lying next to each other with their heads against the midsection of a single large shark. This reminded people of a litter of kittens pushing their heads next to the belly of a mother cat to nurse or drink milk from her multiple teats.

Five days a week, I would push a cart loaded with electronic equipment over to the Shark Channel. On the cart was all of the equipment necessary to project the sounds into the channel and measure their intensity. The playback system was composed of a tape recorder, an amplifier to increase the level of the sound, an attenuator to change the sound level, and an underwater sound projector. The gear used to detect the sounds included a hydrophone (an underwater microphone), another amplifier, and a meter to record the magnitude of the sound. The speaker was suspended in the center of the channel and a hydrophone was placed at a distance of 30 feet from the speaker. I then played back either the scream or a control sound to a shark as it swam under a line passing from one side to the other side of the channel from which the hydrophone was suspended. My control sounds were a single tone matched to the initial *whee* of the scream and a *sheeesh* sound, white noise composed of all of the tones in the scream but with the energy in the sound distributed evenly between the frequencies rather than changing from one tone to another.

To the two sounds from the whale study, a third sound was added, which I expected to attract the sharks to the speaker—the pulsing *borum-bodum-borum.* Day after day, I stood beside the channel from morning to evening playing these sounds to the lemon sharks as they swam by me. Each shark was named for a character in Greek or Roman mythology. The largest male, 8 feet long, was named Hercules; the largest female, 7 feet long, was named Juno. I would increase the loudness of each sound every time a shark failed to react when the sound was played. I kept the level the same if the shark reacted by turning around and swimming away in the opposite direction. I rated the relative effect of each sound by determining the threshold intensity at which sharks reversed direction at the hydrophone three times in a row. Unlike the results of the whale study, there was little difference between the responsiveness of the sharks to the white noise and the scream. The sharks appeared to be startled by the sound, just as you might blink your eyelid when surprised by the flash of a camera or turn your head upon hearing a car screech to a stop. These startle responses are exhibited by all animals, from insects to humans, in response to sudden, unexpected changes in light and sound in their environment.

Consistent with my conclusion was the change that came over sharks when I increased the level of the pulsed low-frequency sounds. My gods and goddesses were attracted to low-intensity sounds, but they swirled around and rushed off when the intensities produced were high. At moderate intensities, the sharks at some times approached the speaker and at other times swam off when the sounds were played. This led me to hypothesize that the sound intensity was critical to the withdrawal response. In fact, it was well known from studies of rats that the rate of sudden increase in the volume of sound was the critical factor in eliciting the withdrawal response. For that reason, my next and final set of experiments compared the response of the sharks to a sound, which was played constantly in their environment, as it increased at different rates. The sharks were frightened when the *sheeeesh* sound was increased only tenfold in the measured intensity (which is equivalent to a doubling in the perceived loudness) and the rate of increase was very rapid. This change in loudness was equivalent to a small but abrupt change in the intensity of your voice when suddenly yelling at someone while playing a game such as basketball or football.

While standing beside the channel hour after hour conducting experiments, I noticed that two or three small, dark brown nurse sharks often swam behind a large light brown female nurse shark. It is easy to distinguish a male from a female shark. Every male has two claspers, conical fleshy protuberances trailing at the side of the pelvic fins that extend outward from the belly just to the rear of the midsection of each shark. The clasper is the shark's version of a human penis. This female differed from the others not only in her large size, but also in the tattered back edge to her side, or pectoral, fins. I brought a pair of binoculars to work and walked, keeping abreast of her as she and her mates swam around the channel together. Sharks normally swim at a speed comparable to that of a human walking slightly faster than usual. Looking through the binoculars, I could see several large crescents composed of small white puncture holes where the teeth of another shark had penetrated her skin. Were these the male love bites that some observers believed were intended to stimulate female sharks to mate with males? These bite marks were only present on female blue sharks and not on males, leading scientists to suggest that they were inflicted during courtship. The female blue shark's back is thicker than the male's back, presumably to absorb this rough treatment.

Yet at the time I stood there watching this trio of nurse sharks swimming around the Shark Channel, no one had yet witnessed the entire sequence of courtship behaviors leading to a male copulating with the female. No more than half a dozen photographs, diagrams, or photographs existed that depicted the male biting the female or inserting his clasper into the vent of the female. Because the male possesses two claspers, an argument raged over whether the male inserted one or two claspers into the vent on the underside of the female.

The bodies of sharks had been examined closely and their internal reproductive system described on the basis of dissections. The clasper, although it resembles the human penis superficially and performs the same function, is quite different in its evolutionary origin and functioning. The organ originates from the shark's right and left pelvic fins, which extend outward at the base of the midsection. The outermost margin of each of these fins is rolled inward into a fleshy scroll with imperceptible boundaries existing between the successive layers of fin. On the end of the organ is a stiff fold of skin with cartilaginous ribs inside, which folds backward against itself. When I wedged my finger between this fold of skin and the rest of the clasper of the blue shark, the fold spread outward like an umbrella. In some species, there are sharp claws at the ends of the cartilaginous ribs. As the umbrella-like structure opens, another stiff fingerlike projection swings outward in some sharks like the spur on a cowboy's boot, and thus is called a spur. These two organs in the clasper penetrate the membrane of the female's vagina during copulation and prevent the male's clasper from slipping out of the female's vent. If you examine the vagina of a female shark after she has recently mated, you will notice that the walls of her vagina will be torn and bleeding after being punctured by these clawlike structures. Ouch!

The male shark has a space between the skin on his belly and the membrane around his body cavity. He fills this space with water. He then forces the water through a valve at the end of this sac into an opening leading to a groove down the length of the clasper. At the same time, the shark releases sperm from sperm sacs, which are located in the belly next to the base of the clasper. By this action, the male inserts packets of sperm into the female's vagina that eventually will burst, releasing the myriad of sperm that will fertilize the female's eggs. Females also have a small organ on the way to the egg-bearing uterus that can actually store this sperm for months, so it can later fertilize her eggs. This may adapt her to make the most of her infrequent encounters with the opposite sex, even when she may actually contain pups (shark babies) in her body from mating earlier with another male and be unable to immediately become pregnant from her most recent sexual encounter. I noticed that in the Shark Channel the male nurse sharks usually swam slightly underneath and behind the females. Male reef sharks had been observed to follow the paths of females even after they disappeared behind coral heads and were unseen by their pursuers. This led to the prevalent idea that males might be attracted to females by a trail of a chemical aphrodisiac, or pheromone, secreted through the vent of the female. But I had never seen shark mating in progress.

And then it happened! I was working late one evening when I saw a male and a female swimming parallel to each other with their tails sweeping synchronously back and forth. Their bodies were on the same level and they were less than two side, or pectoral, fins apart. The male was staying abreast of the female, but occasionally fell back so that he was positioned near her body but just behind her pectoral fin. Perhaps the male was being attracted by a pheromone exuded through her vent, because he was roughly even to the pectorals on her body. I grabbed a small tape recorder on my cart of sound equipment and began to describe what was quickly happening in front of me. After describing the swimming movements of each shark and their orientation to each other, I gave the behavioral pattern a name, *Parallel Swimming*. This was the initial behavior in the courtship ritual and was included first in my ethogram, or catalogue, of the male and female behavioral patterns leading up to copulation.

During courtship, the male nurse shark seizes the female's pectoral fin in his mouth. She responds to this love bite by turning and rolling onto her back. The male then uses his snout to push her into a position parallel to himself. He then swims on top of her so that he can insert his clasper into her vent. (COURTESY OF THE AMERICAN SOCIETY OF ICHTHYOLO-GISTS AND HERPETOLOGISTS)

While swimming slightly behind the female's pectoral fin, the male suddenly dropped his lower jaw and beat his tail quicker, propelling himself forward so that he could now seize the female's pectoral fin with his jaws. I assumed that the male was biting down on the fin of the female at this time because of the tooth marks that I had observed previously on her fin. The large female now pivoted around in front of the male, bending her body into the shape of a boomerang, with her forward torso in front of the male's head and blocking his forward progress. As she pivoted, the female rolled onto her back. I called this behavior *Pivot and Roll*.

The female now lay on her back completely motionless and rigid in front of me, with her back on the concrete bottom of the channel. This was exciting stuff! By this point, I had jumped over the railing at the side of the channel and was holding on by one hand while leaning over the channel to better see what was going on. The male took over now. He pushed the female from a position perpendicular to himself to a position parallel to his body. You can visualize in your mind her body, upside down, being the top of a T and his body the stem of the letter. The male moved her by placing his head and snout between her dorsal fin, which was pointing toward the bottom, and her pectoral fins, which were level with the surface of the water. My name for this behavior was *Nudging*.

Once the female was parallel to him, he swam on top of her and remained there for half a minute. I bent down and looked to either side of the male to see if he had inserted his clasper into the female's vent (also termed the cloaca by scientists), the small opening to her reproductive system, but couldn't see anything because he was between me and the female. Yet now to my surprise, the male rolled onto his back, revealing one of his two claspers inserted in the female. Here was the answer to the one-versus-two-clasper controversy. The single clasper was bent to almost a right angle yet it was staying within the female—a third of it was inside the female. The hook on the end of the male's clasper must have penetrated the vaginal membrane and been holding the clasper inside the female.

I could see the belly of the male now move up and down as he contracted his abdominal muscles to force the water held in his siphon sac through the clasper and into the female's reproductive tract. Both the male's clasper and the female's belly were swollen and tinted red with blood that had flowed to their extremities. I glanced at my watch when the behavior began and ended. The sharks were copulating for two minutes. The two then suddenly righted themselves and rapidly swam off, with the male in hot pursuit of the female. Perhaps they were going to do it again! I felt guilty at the time because I had probably disturbed them, and in my absence they may have mated for even longer. Yet I felt lucky to observe the entire mating ritual of these two nurse sharks. Mating is a rare event in the behavioral repertoire of any animal, be it shark or human, relative to feeding, fighting, or sleeping. Yet during the month of June, many male and female nurse sharks arrive at a shallow water lagoon in the Dry Tortugas, islands west of Key West. Here they perform the same mating ritual. A biologist had observed these reproductive aggregations at the beginning of the last century, and a colleague had given me a series of photos depicting one of these pairs of nurse sharks mating in these shallow waters.

The sequence of courtship behavior is roughly similar among sharks and rays. Even the huge manta ray, which uses large fleshy paddles on either side of the head to funnel small shrimp, jellyfish, larval fish, and other floating animal plankton into its mouth and thus has little need for large teeth, employs biting during mating. The male ray, using his small, nonfunctional teeth, grasps the tip of the huge pectoral fin of the female, a behavior termed *Nipping*. The male then rolls over, while hanging on to the female's fin, so his underside is flush with the underside of the female while he inserts his clasper into her cloaca.

Walking around the channel during spring and summer, I often saw nurse shark mothers swimming around the channel with the tail or head of a baby hanging out of her vent, the consequence of mating a year earlier. The babies would shake their bodies back and forth to dislodge themselves from mom's vent, then dart off in search of a hiding place in the channel. From the side of the channel, they looked like miniature replicas of the adults. The flat, featureless concrete bottom of the channel afforded little shelter. Unfortunately, the other sharks in the channel, the lemons and an occasional large, 12-foot-long tiger shark, would eat them. One day, I couldn't restrain myself from scooping one up in a dip net to keep in an aquarium above my desk at the marine laboratory. I was careful when removing the little nurse shark from the net. There is a correct method of holding small sharks both under and out of water. You do not grab the shark by its tail. It may quickly pivot around and give you a nasty bite. Rather, I gripped the shark with two hands, one just behind the gill slits and the other just forward of the base of the tail. The baby shark that I held in my hands was very attractive. It was brown with alternating dark and light bands, each of which merged into the other with a gradual shift in the hue of its color. Speckling his body were small green iridescent spots. Both the bands and spots would gradually disappear as he grew up. Protruding from the forward edge of his nostrils were long barbels. These gave this foot-long nurse shark, whom I named Huey, a comical touch.

The nurse shark is a sluggish animal, preferring to lie on the bottom. This trait has lead to its frequent harassment by divers, whose unwanted attentions have, at times, provoked this normally mild-mannered shark to bite humans. The sedentary character of the species makes it possible to keep nurse sharks in small aquariums. Because I had seen nurse sharks partly hidden under ledges on coral reefs, I provided Huey with a piece of drain pipe for shelter in the aquarium. He often swam into it and remained there, tail and head protruding. At dinnertime, Huey was not at all sluggish. When I dropped a shrimp on the surface of the aquarium, he would promptly swim up to the shrimp, open his mouth, and expand his large throat cavity, sucking in the shrimp with a strong inward movement of water that resulted in a loud slurping sound.

The public's image of sharks as mindless is reinforced by the way their behavior has been interpreted and misunderstood. For example, sharks are often said to be indiscriminate feeders. An early study involving dissections of tiger sharks did find an odd assortment of items in their stomachs, ranging from beer cans to shoes. However, this apparently omnivorous feeding behavior just might be explained by the fact that the sharks were caught near a municipal garbage disposal site where such items were dumped into the water. The sharks may have ingested such objects because they need to use the garbage as a kind of "ballast" that keeps them from being so light that they float or so heavy that they sink. College anatomy courses further portray the shark as a simple animal—the shark dissection is used to represent a "primitive" vertebrate, while the cat dissection denotes an "advanced" vertebrate.

I wondered whether Huey was really a dumb animal. The willingness of both juvenile and adult sharks to learn had been used by behavioral scientists to describe the sensory abilities of sharks. For example, a learning technique had been used to establish the shark's sensitivity to sounds and the limits of its ability to distinguish a more intense sound from another sound or a tone of a certain pitch from another tone of similar pitch. Sharks like Huey are very sensitive to the low-frequency sounds produced by their prey. These tones may be so low that we can't hear them, even though they sound very loud to a shark. But sharks are not quite as good as we are in discriminating between sounds of different intensity and pitch.

The method by which learning is used to investigate the sensory capacities of the shark is simple. A stimulus (termed reinforcement) such as a mild shock is given to the animal to produce a reflexive response such as a blink of the sharks nictitating membrane, the eyelidlike organ of the shark. A second stimulus is applied to the animal either prior to, or at the same time as, the shock. This might be a tone of a certain pitch within the hearing range of the shark or a light of a color within its color range. After a number of shocks are coupled with the tone or light, the shark will blink its nictitating membrane in response to the tone or light without the shock being applied. The pitch of the tone or the color of the light can then be varied, and the shark's perception of them is indicated by whether it blinks after the stimulus is presented.

Alternately, reinforcement such as food or a shock can be presented to the animal immediately following a certain behavior to induce it to learn. After pairing the stimulus with the response a number of times, the frequency of the behavior's occurrence will increase if the reinforcement is positive (as when feeding) or decrease if the reinforcement is negative (as when shocking the shark). If these reinforcements are discontinued, the learned behavior will become less frequent.

I set out to train Huey, the nurse shark, to push his snout against a white target. First, I had to make the target. I constructed this of a small, white Plexiglas square, 3 inches by 3 inches, which was attached to a narrow strip of wood. I made the strip of wood long enough so that I could position the target close to the bottom of the tank where Huey swam while keeping my hands out of the water. I had to be careful because after he learned to bump the target in quest for food, he occasionally mistook my hand, whitish in color underwater, for the target and attempted to take a bite out of it.

Ideally, my subject should be given a food reward only after it bumped the target by chance while swimming around the tank. In practice, I found this approach time-consuming. To speed up the learning process, I used a technique referred to by behavioral scientists as "shaping." Huey was accustomed to being fed pieces of shrimp or fish with a pair of salad tongs. I fed him three or four pieces every day. After he became used to being fed in this manner, I lowered the target to an upright position close to the bottom in front of him and held the food right in front of the target. He swam toward the food, swallowed it, and kept moving forward until he bumped into the target. Sometimes, he would not do this, so I carefully held food close to him and he followed the food to the target. I was careful to make sure he hit the target as he swallowed the food. Finally, a few times, I let him grasp a piece of fish held in my hand, and dragged him, hanging on to the fish, to the target. After fewer than a dozen times, Huey associated bumping the square with a food reward. To find out whether he had made the association, I now presented the target without any reinforcement or shrimp present. Aha! He punched the target with his snout.

Huey was succeeding well in the learning trials, but he had grown half a foot longer. The size of the aquarium will influence the rate at which a fish grows larger along with the amount of food that it is fed. A fish that is confined in an unnaturally small container can secrete a chemical that inhibits growth. For this reason, I moved Huey to a larger tank, 4 feet square, at the laboratory, where the rest of his training could be completed.

The next step was to teach Huey to bump one target and not a slightly dissimilar one. To do this, I presented him with two targets, the white one and a black one of a similar size and square shape. I did this in alternating fashion and fed him after he bumped either target with his snout. Huey always pushed the white target, but only occasionally the black one. He chose the second target only because he had associated his reward (food) with not only the color of a paddle but also its size and shape. I then fed or reinforced him only when he bumped the black target and not the white one. After a while, Huey struck only the black target and not the white one. Feeding wasn't even necessary to get him to do this. This type of learning is called a "discrimination" because the shark learns to choose one target over another on the basis of a difference between them. This task is more difficult to learn than the one described earlier. If I had continued to run my school for sharks, I could have taught Huey any number of discrimination tasks. He would have been able to distinguish between targets of different sizes, shapes, textures, colors, and color patterns. However, I stopped training Huey at this time. Being a voracious feeder, he had now grown 3 1/2 feet long and was fast becoming too large for his third aquarium. I was happy to release Huey into his natural environment, the shallow waters surrounding the Miami Seaquarium, but I suspected that the lessons that I had taught him were not going to help him become a successful predator in his natural environment. Nurse sharks feed not only on shrimp, but also on spiny sea urchins, lobsters, squid, and large shellfish such as the queen conch. When hunting the conch, the shark turns them over and somehow manages to extract the living animal. This must take some intelligence because the conch's soft body is retracted into a scrolled shell and it wedges the hard plate attached to its foot into the small opening of its hard shell to block access by the predator.

These simple experiments convinced me that sharks are intelligent. Of course, this was prior to my discovering the variety of their behaviors and complexity of their social relationships in the ocean. And how do sharks compare in learning ability to other animals? Eugenie Clark, the first women scientist to work with sharks, compared the ability of lemon sharks to learn to capabilities of goldfish and mice. The sharks learned more rapidly than goldfish and at a similar rate to mice. That's not bad for supposedly dumb feeding machines.

One day Art and I were together on a research expedition to the Tongue of the Ocean, playing back sounds to silky and whitetip sharks. The seas were rough, sharks were swimming everywhere, and we were trying hard to avoid being bitten. Both of us were a little sea-sick. We dragged ourselves out of the water onto the deck of the boat and lay there, breathing heavily, because we were exhausted by the ordeal of working in these marginal conditions. Art looked at me seri-ously, and said: "This type of work is for young men, not for a middle-aged man such as me. I've had two heart attacks due to the pressures of this arduous work, and the Navy is never totally pleased with our performance. I'm going to concentrate on my studies of small dam-selfish in the local reefs and stop studying sharks."

This decision forced me to look elsewhere for an opportunity to get my Ph.D. learning about the behavior of sharks. I immediately contacted Donald Nelson, who was a professor on the faculty of the California State University at Long Beach. Our interests in shark behavior were mutual, and he currently employed students on a grant from the Office of Naval Research to study the field behavior of sharks. However, his institution awarded only master's, not doctoral, degrees. I also talked with Richard Rosenblatt, one of the premier ichthyologists (fish biologists) in the United States about working under his supervision on a project on sharks. Dick, who was a won-derful teacher with an encyclopedic knowledge of fish biology, was a professor on the faculty of the Scripps Institution of Oceanography, a research unit of the University of California at San Diego. This insti-tution did confer doctoral degrees, but to be accepted in the graduate program, one needed not only stellar grades, but also to provide assurance that adequate financial backing would be forthcoming to support one's studies and research. I asked Dick and Don whether they might supervise my doctoral studies jointly. Dick would chair my doctoral committee, which would be composed of prominent scientists from Scripps along with Don (despite his being affiliated with another institution), yet I would work and be paid through Don's grant from the Office of Naval Research. The funds for my salary would have to be transferred from California State University to the University of California.

Don discussed this arrangement with Eric Shullenberg, who was the program director of the Division of Oceanic Biology at the Office of Naval Research and a recent graduate of Scripps. He was interested in this proposition because he had been pleased with my performance while working on Art's grant to study the acoustic behavior of sharks. The Navy has always had a commitment to basic research, and to enabling students to complete advanced degrees in areas relevant to the Navy's interests. It was one Friday afternoon during the spring of 1977 at the TGIF (Thank God It's Friday) party in the lounge of the University of Miami marine laboratory that I felt a hand on my shoulder. I looked around, and there standing next to me and smiling was a young man about my age, who introduced himself as Eric Shullenberg. He said to me, "I've got good news for you, Pete. Yesterday, we decided to fund your studies in California." I shouted with joy.

And so in the summer of 1977 I moved from the University of Miami, where I studied for my masters degree since 1973, to the Scripps Institution of Oceanography in San Diego, to work on my Ph.D. I chose to study hammerhead sharks, a Ph.D among sharks, equipped with a large brain, rich and complex social behavior, and, as we will see, an amazing ability to navigate at sea. The hammerhead has long been my favorite shark, and it is more representative of sharks generally than the better-known white shark, made famous by *Jaws*. (Physically more impressive, the white shark is the athlete of the shark world.)

Moving to Scripps gave me the opportunity to dive down into the middle of schools of hammerhead sharks and unravel the social secrets of their societies at a seamount in the Gulf of California. Because I knew so little initially about the behavior of hammerheads, I had no clear idea of how much danger I would be in when I approached them. I admit that a certain high-pitched level of anxiety was vibrating in my mind. But this was the only way to learn more—to enter their world and risk what might happen. Later I would study white sharks. But the hammerhead provided my first view of the hidden world of sharks—the one that can't be captured through the bars of a shark cage. I was both excited and apprehensive on the day I first slipped into the blue water of the gulf and plunged into the depths of the shark's kingdom.

CHAPTER 4

DIVING WITH SHARKS IN SOUTHERN CALIFORNIA

The Scripps Institution of Oceanography is an immense marine laboratory with numerous buildings situated along a cliff overlooking the white sandy beaches and clear blue waters of La Jolla in Southern California. I entered the Ph.D. program at the graduate school during the fall of 1977 and was working half-time on Don's grant from the Office of Naval Research to conduct field studies of sharks. The Navy wanted us to develop countermeasures to reduce the hazard sharks posed to sailors and airmen lost at sea, but first we needed to gain a basic understanding of the behavioral habits of sharks. I was now keen to observe more shark behavior in their underwater world, but just how to do this was not at all obvious to me. I was well aware that the first, and often the most difficult, challenge to an ethologist was to make reliable contact with the subject of his or her study. This was to be my first goal while a graduate student at Scripps. Not only did I need to find a place inhabited by not one or two sharks but large aggregations of them, but I also needed to find sharks that would stay put long enough to allow me to observe them repeatedly. I would have to approach them closely underwater to see their behavioral patterns and note their social relationships.

The first species that I attempted to study underwater was the leopard shark, a small species with the coloration of a leopard that is common in shallow water close to shore in Southern California. Since I experienced little success in swimming among schools of this species (they fled upon being approached), I turned my attention to the blue shark. This is a graceful and slender gray shark with a bluish hue, which is found several miles from the coast of Southern California and is common at temperate and subtropical latitudes across the Pacific and Atlantic Oceans. To attract these oceanic sharks to the boat, I had to place chum in the water. This caused the sharks' actions around the boat to be confined to simple feeding responses. Because these sharks were so preoccupied with feeding, they did not exhibit the complex behaviors that I sought to study.

These first encounters with sharks off California were not a waste of my time. These experiences led me to wonder why sharks performed odd and unexpected behaviors such as basking at the surface and biting metal parts of the cage that protected me while viewing them. I also discovered that leopard sharks swam in schools and noticed that several blue sharks often arrived at nearly the same time when I was attracting them with chum, suggesting that they might also form schools. This led me to question whether sharks were social species and not solitary as scientists believed at the time. Furthermore, I was forced to develop techniques for observing sharks unobtrusively, marking them to keep track of individuals, and statistically inferring that they did not arrive randomly at the source of chum.

Not long after arriving at Scripps, I was called by a fishery biologist who excitedly told me that there was a big school of sharks in the shallow water off the beach below his office. He worked at the Southwest Fisheries Science Center, a facility of the U.S. National Marine Fisheries Service which was a three-story building perched high on the cliffs overlooking the next beach north of Scripps. I grabbed my camera and hurriedly climbed up the hill to the Fisheries Center less than a quarter of a mile from my office. The biologist was standing outside his office on the second floor of the building next to a telescope. This telescope was used to record the passage of dolphins and whales (and also to view bathers on the beach, which because of a rock cliff preventing easy access, was home to a large nudist colony). It was a sunny day early in the third week of April; the waves coming ashore were widely separated and small, and the water was crystal-clear. There below were the dark outlines of many sharks slowly circling with their tails slowly moving back and forth in the light blue-green water below. At this distance, they looked small and, with their rounded bodies and short tails, resembled larval frogs or pollywogs in a puddle of water. (I could not identify the type of shark because they were too distant.) Counting the number of sharks in a group is always an imprecise task because one can often count the same individual twice, especially when you cannot see the entire school at once. My usual solution was to count the members of a school three times and compute the average. I counted 27, 24, and 30 individuals, which gave me an average of 27 sharks in the school. I took half a dozen pictures with my camera using a telescopic lens, which at my distance viewed an area the size of half a football field, so that I could make a more accurate shark count later.

I returned to the Fisheries Center on the following day to make an ethogram consisting of descriptions of the behaviors performed by the sharks. To start with, this was a very curious grouping. The sharks were not moving in separate directions, but were combining into formations. Half a dozen of the sharks would form a line, the second shark one body length away from the leader, the third a similar distance from the second, and so forth. These formations often proceeded in one direction along a sinuous path. There were at least four or five of these parades going on at any time. The leader of the formation often slowly turned until it came up behind the last shark to form a large circle with all of the sharks swimming in the same direction. At other times, the members of one line swam by the members of another line, moving in opposite directions. As each shark in one line passed, it often turned its head toward the midsection of the other shark passing in the opposite direction. I named this last behavioral pattern *Tandem Swimming in Opposite Directions* to differentiate it from the more common *Straight-Line* and *Circular Tandem Swimming* that I had observed that day. This pattern reminded me of the three male nurse sharks constantly swimming alongside a female nurse shark and occasionally turning their heads toward her midsection to smell her pheromone, or chemical aphrodisiac. Single sharks also often nudged the midsection of others when approaching closely as though they might also be checking out their sexual readiness. Was this the same mating ritual practiced by the nurse sharks?

The ritual reminded me of a human custom that I had witnessed a few years earlier in a small town in the Mexican province of Veracruz. The bachelors, talking among each other, casually strolled around the town square in one direction, and the women eligible for marriage walked together in small groups in the opposite direction. As a girl passed a group of men, she might single out an individual whom she liked, look at him, and smile. If he liked her, he would respond to her demure advance with a wide grin of his own. This was often the first step to human pair-bonding in Mexico.

Probably the most enigmatic shark behavior of all was what I called *Ventral Body Upward Movement*. A slowly moving shark would suddenly speed up, roll over onto its back, and right itself after a second or so. Sunlight reflected off the shark's white belly, producing a bright flash of light that I could easily see even from my distant vantage point on the second-floor balcony of the Fisheries Center. Single sharks performed this behavior most often, but there were exceptions to the rule. For example, each of the sharks in a straight-line formation would roll onto its back when it reached the same spot near the rocky reef. To this day, I do not know the function of this complex and intriguing social pattern.

I had been observing the behavior of these sharks for two days and still had no idea which species they belonged to. My office mate, a native Californian, suggested that they were leopard sharks because members of this species commonly lived near rocky reefs. Here their mottled black and white appearance matched the dark rocky bottom and provided them with camouflage. This was a relatively small shark common in Southern California. Adults were no more than 4 to 5 feet long. I needed to get close enough to these distant shapes to see if they were really leopard sharks. The sharks were still there during the morning a week later. I drove over to the area in a boat, anchored it just beyond where the waves were breaking, and swam toward the site alone, wearing a mask and snorkel. I moved slowly, hardly kicking at all, in order to approach the members of the school without frightening them. After coming to the rocky reef and barely recognizing three mottled leopard sharks in a procession barely visible in the distance, I heard a thunderous *varooom* caused by the sudden agitation of a great amount of water. The three sharks and others I couldn't see had detected my presence and suddenly darted off explosively in all directions, leaving not a single shark in the area for me to observe. This was discouraging to an aspiring shark ethologist. I had won the battle, identifying the species, but lost the war because members of the species appeared to be unapproachable in the water.

Their sudden evacuation resembled their coordinated response to a pod of pilot whales, each 10 to 15 feet long and passing less than 25 yards from them, that I had observed a week earlier. The members of the school darted off in all directions, while suspending sand in the water and reducing the clarity of the water in the process. Perhaps the sharks favored this shallow-water home or habitat to avoid being discovered by large predators. The surf zone was surely a safe place: bubbles and sand suspended by the breaking waves would hide them from the sight of predators and the noise produced would mask any sounds their chafing bodies made on the sand and rocks. Why would the sharks flee from pilot whales? Perhaps they mistook these intermediate-sized whales, which generally feed on fish and squid, for larger killer whales, which might prey upon them. Alternatively, there are reports of bottlenose dolphins dashing toward sharks and butting them with their thick fleshy rostrums. This has usually been interpreted as a defensive tactic used to drive away predators, but also could alternatively function to drive away competitors from an area containing a mutually desirable resource. Indeed, leopard sharks feed at times on small fish as do dolphins—and some of these small fish are present in the surf zone. In short, there is an abrasive relationship among dolphins and sharks. In fact, the school members would be a good deal safer if they responded to any intrusion by another species into their private space. I had to find another shark for my behavioral studies, one that would let me approach it more closely and stay with it longer. In addition, these sharks left the area after spending only ten days below the Fisheries Center. I needed sharks that would stay around longer than that to learn about their behavior.

It was precisely a year and three days later that I received another telephone call from the same biologist saying that the leopard sharks were back in exactly the same spot below his office. They had arrived with clocklike precision, but again stayed for less than two weeks. They were supplanted by a huge school of roughly a hundred guitarfish, a sand-colored, flat-bodied shark whose forward body resembles the shape of guitar. The shark's snout is pointed and its flattened body expands laterally into a triangular midsection, which is formed by two large pectoral fins that are held close to its torso. Its two smaller pelvic fins extend outward from the constriction at the base of the pectorals to form a second triangle facing in the opposite direction. This double expansion of the body, separated by a narrow waist, resembles the resonant chamber of a guitar, while the shark's long tail looks like the neck of the guitar. All of these sharks lay on the bottom with their heads facing incoming waves. Unlike the leopard sharks, the guitar sharks performed very few behaviors while grouped. The sharks lay on the bottom hour after hour, no more than half a body length from each other, but showed little interest in their neighbors. The guitarfish, like the leopard shark, did not appear to be a good subject for my doctoral studies—they were too inactive.

My observations of the leopard sharks did convince me that some sharks were not solitary hunters that punctuated periods of inactivity with voracious feeding as was the public's perception. These sharks were not solitary at all, but joined groups and performed intricate maneuvers with other members of the group. Were these sharks performing a mating dance that paired males with females? As much as I wanted to understand the function of the dance, I would never be able to do so unless I could easily approach them in the water. How else could I see the claspers on their underside and thus distinguish males from females? My distant hillside observations of the leopard shark were never going to unravel the secrets of shark behavior. I had to find another subject for my study, a species that could be observed underwater. This meant I would have to find a way to make sharks come to me. Chumming had been used to attract blue sharks, which are common off both the eastern and western coasts of North America. Why not observe blue sharks attracted by a mixture of ground fish—but not wait for them in a shark cage? I needed to see them but at the same time didn't want them to see me. A shark cage wasn't a very good vantage point from which to observe the sharks, because they could see the person inside, and might pay more attention to this person than to the other sharks with them. I needed to find some other method of observing sharks, while staying out of sight.

A common practice among biologists studying sea lions was to view them from a small hut—a blind—situated at the colony. Whenever a human walked on a beach with sea lions, the members of the colony would begin to bark loudly and then stampede into the sea. The solution to this problem was to build a small wooden building at the colony with a narrow opening, from which the person could see the sea lions but they couldn't see the observer. The sea lions quickly became accustomed to this nonmoving, inanimate object, which was now accepted as a constant part of the natural environment. My blind was a floating platform, which consisted of a flat sheet of 4-by-8-foot plywood surrounded by 2-by-6-inch wooden sides, to which were fastened half-cylinders of foam flotation material. The bottom was tapered toward one end to fit snugly in the pointed bow of my boat and had a 2-foot-wide square notch in the front, through which I could squeeze my chest and head in order to look downward into the water.

Twice a week, Jeremiah Sullivan and I would roll the trailer with my boat to the end of the Scripps Pier and lower it into the waters below. Jeremiah was the son of a motion-picture actor and fitted that role perfectly; he was slim, muscular, and handsome. He was attending the Extension Program at the University of California and was keen to learn more about the behavior of sharks. He wanted to become an underwater cinematographer, and the shark was a popular subject of underwater films. The pier ran a couple of hundred yards out from the cliff and over the water, extending beyond the surf zone, where the waves broke. Here the water was 20 to 30 feet deep. The wooden deck of the pier was 20 feet above the water and was supported by massive cylindrical pilings made of concrete that jutted out of the water.

Once at the end of the pier, we clipped two large pelican-bill-shaped hooks onto metal rings on either side toward the rear of the boat and one to a ring at the front of the boat. Rope lines led from the hooks to a large plum-shaped metal ring. One of us stood up on the middle seat of the boat and clipped this ring to another larger pelican hook at the end of a stout metallic cable lowered from a large beam that crossed the pier and extended over the edge. The other pressed one switch, that turned on an electrical winch, which lifted the boat up and off the trailer, and then pressed a second switch that slowly moved the boat to the side over the edge of the pier. The first switch was then activated to lower the boat downward into the water. One of us quickly climbed down a ladder holding a line leading to the boat. Although waves rarely broke here, the swell was often large. The boat could move up 5 feet then down another 5 feet with the slowly oscil-lating surface of the sea. It was important to hold the pointed bow in the direction of the incoming waves so that the wave crest did not hit the side of the boat and roll it over. Both of us would then jump into the boat, turn on the engine, unclip the harness, and speed away from the pier. Speed was critical here because the boat could capsize if it were hit by a large wave that happened to break farther than usual from shore. However, with practice both of us worked quickly and could drop the boat into the water and get away from the pier in a matter of seconds.

We next drove the boat 3 or 4 miles offshore and began chumming for sharks. Along the way, we often saw blue sharks and an occasional mako slowly swimming at the surface. They seemed to swim endlessly in a straight line parallel to the coastline, yet they were too far from shore to use the shoreline to guide their directional movements. These sharks tend to swim at the surface of the water, behavior that was to eventually be the focus of my research studies. Why swim at the sea surface when your gills, fins, and scales are adapted for life underwater? There surely must be some benefit to the shark from this inexplicable behavior? And how did they manage to swim in a straight line without seeing some guiding feature, such as a ridge on the bottom.

Some scientists have suggested that sharks can better see the sun when they swim at the surface and can use it as a kind of navigation aid to help them swim in a straight line. Others have argued that they swim at the surface of the ocean because this is where the earth's main magnetic field is most uniform, which helps to guide them in a uniform direction. This takes a bit of explanation to understand. The main field is the portion of the earth's field that resembles a large magnet with positive and negative poles. These magnetic poles are situated near the north and south geographic poles, around which the earth rotates once each day. If you pass a copper wire, which easily conducts electricity, through a strong dipolar (two pole) magnetic field, a flow of electrons—or current—is induced to pass through the wire from a source of the electrons to their destination. Every element, even a good conductor such as copper, has some resistance to electron flow. Hence there will be more electrons at the source end of the wire than at the destination end, a gradient in the concentration of electrons—or the voltage. The magnitude of the current and voltage depends on the conductor's direction of movement relative to the axis between the positive and negative poles of the magnetic field.

Sharks were known at this time to be capable of perceiving the minute electrical field created as they swim though the earth's main magnetic field. They have organs sensitive to electrical voltages, the ampullae of Lorenzini. These organs consist of pores on the underside of their snout around the mouth that open from spaghetti-like canals filled with a gelatinous material, which is a viscous solution of charged ions that slowly conducts electricity. There are fingerlike protuberances at the bases of these canals that are sensitive to the gradient in electrons across the canals—or voltage—induced by the shark's movement. The magnitude of this personal field varies with the direction of the shark's movement relative to the axis between the earth's north and south magnetic poles. The shark can hold a steady course by avoiding turning and changing the personal voltage induced through these organs. This is easiest to accomplish where the main field is a uniform dipole, undistorted by other sources of magnetic fields. I will talk more about these organs later and explain another way hammerhead sharks may navigate.

We usually stopped once we were far enough from shore that it was barely visible on the horizon. Then we would throw a plastic cylinder into the water. Openings in the container were covered by screening that allowed blood and body fluids from ground-up mackerel inside to flow out. Once we had our version of chum sending out an attractive scent, we would lift the observation platform onto the side of the boat and slide it into the water. I then put on my wet suit, pulled the platform near the boat using a rope attached to its rear end, stepped out of the boat, and crawled on top of the platform. On most days, the sea surface consisted of large rolling waves with rounded peaks and valleys. Sailors call this the swell, and these waves are created by strong winds hundreds of miles away. The platform would tilt slowly as it rolled in the swell. It was the choppy seas formed by strong local winds that rocked and jerked the platform back and forth. I would hold on to its sides with all my strength to avoid being thrown into the water with the sharks circling the platform. Even a longtime sailor such as myself had occasional moments of sickness. You just couldn't help it. After vomiting (and adding beneficial chum to the water), I always felt better and returned to observing the sharks.

The blue sharks arrived, usually one by one but sometimes in twos and threes, and began to swim around the platform. They initially swam in that slow, sinuous manner that reminded me of the way snakes move on land. What elegant creatures, with their slender bodies and long and curved pectoral fins! Often they would quickly beat their tails back and forth two or three times and then hold their tails relatively still while gliding effortlessly forward, using their long upwardly tilted pectoral fins to propel them toward the surface.

As more and more blue sharks arrived, however, their behavior changed in a radical and frightening way. They began to beat their tails faster, bending their bodies into right and left angles, and darted off to the side or even spun around to swim in the opposite direction. The water became full of sharks frantically darting here and there. They were like a hysterical crowd of dancers moving back and forth on the floor. On one of our first days at sea, one shark dashed toward the front of the platform, from which my head and arms were protruding. I quickly withdrew my hands and arms from the water, and the shark passed within a foot or so of my head while opening its jaws. The shark then grabbed the Styrofoam on the side of the platform with the pointed teeth of its lower jaw, and in an instant moved its upper jaw back and forth with a sawing motion, to remove a crescent of soft plastic from the platform. Had the shark intended to bite me? This was a sobering sight that sent a shiver down my spine.

More than once during our trips when I lay on the platform, sharks that were not in my field of vision would surprise me by bumping into my arms. It was simply impossible to keep track of all of the sharks circling the platform at once and pull my arms out of the water whenever one approached to bite me. For this reason, we installed a cage on the platform to protect me. This consisted of a square box that fit over the notch for viewing at the front of the platform. It was composed of wire bars the width of a pencil with a two-inch separation between each and offered me ample protection.

However, the blue sharks did often seize the metal bars of the cage. They did not seem intent on biting me but were attacking the cage itself. It had been dipped in a vat of molten zinc to reduce corrosion. The bars were now composed of two metals, iron and zinc, that had different electronic charges in salt water, which is a conductor. Electrons flowed between the zinc and any uncovered iron like the current flowing between the plates of a lead-acid battery, creating an electronic field, to which the blue sharks were sensitive. Could the blue sharks be mistaking the fields of the bars for those of their prey? Other shark species that inhabit shallow water (such as the juvenile scalloped hammerhead shark) locate creatures they prey on even when the prey are buried and out of sight. For example, the wrasses, a group of small, cigar-shaped fishes inhabiting seas in the tropics, bury in the sand for safety at night but are given away because they generate minute electromagnetic fields that are detected by juvenile hammerhead sharks as they patrol back and forth over the bottom in search of a meal. The hammerhead has a laterally enlarged head, partly to increase the width of the path that it patrols. Evolution has favored a laterally enlarged head for the same reason that a human engineer chose a flat, rounded disk for the shape of a metal detector designed to efficiently find buried coins. When aroused by chum, the blue sharks appear to become less selective in their feeding behavior and bite even inanimate objects that emit electrical fields.

The blue sharks didn't perform many different behaviors when chummed. They often accelerated suddenly without any obvious provocation. When approaching the bait container, they often rolled onto their sides and rubbed their bodies against it as they passed. At times, the sharks would quickly turn around and follow the next shark approaching the source of the bait. Smaller sharks seemed to move to the side when a larger shark approached. There just wasn't much behavioral variety near the bait bucket. The sharks were there to feed and paid little attention to each other. Only when I left the confines of the platform to photograph the sharks did they exhibit two defensive behaviors. A shark accelerated toward me shaking its head; another swam in a stiff manner to the side of me with lowered pectoral fins. This was likely a threat behavior. Here I was able get a closer view of the sharks, but the blue sharks, unlike the leopard sharks, exhibited only simple behaviors. I wasn't learning much from observing the behavior of sharks attracted by food in the water. All you got was biting, biting, and more biting. It was imperative that I study sharks that were not being fed by humans.

A practical problem at this time was how to distinguish among sharks. There could be between ten and twenty sharks swimming in separate directions around the platform at any one time. How could you know when a new shark arrived? There were so many others of similar size and the same sex. I needed to mark those already present in order to distinguish them from newcomers. I tried many methods of marking individuals. The easiest and most effective was to use a pole spear to insert a tiny metallic barb into the back of the shark with an attached length of monofilament line threaded with colored vinyl tubing. Using four or five colors, one could create many unique combinations, such as white, blue, yellow, and red, which could easily be distinguished from red, yellow, white, and blue.

I mostly used these spaghetti-like tags to mark sharks, but tried to develop a tag that would last a single day on the shark. I first inserted a metallic dart into the shark with several Life Saver candies of different colors attached, but they dissolved before the end of the day. Also, it bothered me that the dart might hurt the shark. So I started attaching stainless steel alligator clips covered with vinyl of different colors to the dorsal, or top, fins of the sharks. This clip was a metallic version of a clothespin, with serrated metallic jaws instead of smooth wooden

pins forced together by a spring. One pressed down on the ends opposite to the jaws to open the clip. I reached over the platform as a shark
passed with the clip in my hand, opened its jaws by pressing my fingers together, and let go so that the jaws came together on the soft
dorsal fin of the shark. Jeremiah also pulled the bait canister to the
side of the boat so that the sharks swam close to the boat, where we
could attach the alligator clips.

The motivation of these sharks to feed was so heightened by our
putting the smell of fish into the water that the sharks were less selective about what they tried to eat. Neither the platform nor the cage
resembled their usual prey—small fish and squid. This seemed odd
behavior to me. They were emboldened by the presence of many others. This motivated them to feed on an item in their environment that
looked different and was larger than their usual meal. Behavioral scientists refer to the more direct and brazen behavior of animals in the
company of others as "social facilitation." This is common among
members of the dog family, such as wolves that hunt within groups
and act in concert to kill prey larger than themselves—for example,
moose. Fisherman and divers called this type of behavior among sharks
a "feeding frenzy." While the wolves moved as packs and captured
their prey by their coordinated action, the sharks appeared to act
independently of each other when aggregating at the bait bucket.

Could blue sharks form and move in packs like wolves? This question crossed my mind one day as we drove the boat back to the Scripps
Pier. It occurred to me to examine the passage of time between the
instant when each shark arrived at the boat. If the sharks formed packs,
five or six sharks should arrive immediately, one after another, and then
there should be a long hiatus before another group arrived. If the sharks
were solitary, they should arrive after long periods of time, all of a similar duration.

It was essential to know the proportion of short to long intervals if
the sharks were distributed randomly throughout the ocean like a
handful of marbles scattered on the ground. When I explained this
new idea to Dave Lange, the mathematical guru of Scripps, he quickly
replied that my question was an old one. This brilliant guy impressed
all of us with his amazing knowledge of mathematics and ability to

use that knowledge to answer real questions about the distribution of animals in the sea. In class, he stood in front of us and derived complex statistical and mathematical formulas from his head, a task that even the brightest among us had yet to master. He went on to say that there was an arcane but important branch of mathematics, called queuing statistics, that was developed to answer my specific question. These statistics were used during World War II to determine whether German submarines should search for Allied convoys alone or in "wolf packs." The same mathematical method enabled banks to match the number of tellers on duty at a window to the varying rates at which customers arrived throughout the day. There were many people who withdrew money at lunch, yet fewer during work hours. These were useful mathematics, which could be applied to submarines, bank tellers, and blue sharks.

Dave asked me to come by his office during the next day and he would have the mathematics worked out for me by then. When I passed him in the hall on the next day, he drew out of his pocket two pages of paper with an impressive title scribbled on the top of one: "Derivation of the Exponential Distribution of Spacings in a Poisson Process." On these pages were a series of formulae, including integrals and derivatives from calculus, that led to one simple formula. This could be used to calculate the number of intervals of different lengths of time you might expect if the sharks were scattered randomly at sea.

My next step was to separate the intervals between the shark arrivals into a series of classes—say, the most frequent arrivals would occur with intervals of 0 (two sharks arriving simultaneously) to 4.9 minutes, the next from 5.0 to 9.9 minutes, and so forth. I then used Dave's mathematical formula, entering my average interval between shark arrivals into it, to predict the number of intervals that would occur in each of the classes of time, if the sharks were randomly spaced. I was really surprised! The random distribution contained a lot of very short intervals and only a few long intervals. Things happen randomly soon after each other. One has always heard that bad things occur in threes or fours, perhaps a result of this unexpected outcome. Before, I had the impression that the blue sharks were forming groups, because after a period of absence they all appeared at once. However, now there was little difference between my large number of short intervals between blue shark arrivals and the number of short intervals predicted by Dave's mathematical model for randomly spaced sharks. The K-S test, a statistical procedure named after the Russian statisticians Kolmogorov and Smirnov, indicated that there was no significant difference between the intervals at which my blue sharks arrived and those for sharks swimming alone in the ocean.

It was while studying blue sharks in the waters off Scripps that I escaped one of the greatest perils of my career. The risk came not from being attacked by sharks but from being in a boat in bad weather. A high school biology teacher and one of his students, who was conducting an experiment comparing the relative importance of vision, smell, and hearing in blue shark feeding, accompanied me on one of my routine trips out to an area where sharks were plentiful. The experiment was for a project that was entered into a competition in the student's high school science fair. We were outward bound when a furious thunderstorm, accompanied by howling wind and huge breaking waves, broke out without any advance warning. I quickly brought the boat back to the Scripps Pier and came as close to it as I dared. I hurriedly told my guests to tighten up their life vests and swim over to the ladder leading up to the safety of the pier. There was no way we could ever lift the boat out of the water with waves crashing down even at the end of the pier. They might not be able to swim to shore if the boat were sunk far offshore. As an accomplished competitive swimmer, I was confident of my ability to reach shore. I told them to call my wife, who would tell them where to drive my van and meet me in Mission Bay, a harbor 10 miles up the coast. I contacted the Coast Guard over the boat's radio and informed them of my plans, then began to drive the boat along the coast, just outside the surf zone, where huge waves were breaking on the cliffs of La Jolla. I could swim to shore at this close distance if the boat were capsized by a breaking wave. It was dark by the time I entered Mission Bay in my boat, exhausted yet determined to find the boat ramp, where my guests would be waiting. What a relief when I finally spotted them, waving their flashlights excitedly in the distance. I winched the boat up onto my trailer, which they had brought to the site with my van, secured the boat tightly, and then got in the van and drove off. When I got to my office at Scripps, the phone rang. It was my wife. She asked me, in an incredulous tone, "Don't you notice something missing?" I said "No," to which she replied that in my haste I had left my guests standing on the boat ramp, shouting loudly and waving their arms frantically, trying to get my attention as I drove off without them.

It was soon after this brush with misfortune that I decided that neither leopard nor blue sharks were ideal subjects for my studies. My attention was soon to be drawn to scalloped hammerhead sharks, which were rumored to assemble in huge schools at underwater mountains (seamounts) in the Gulf of California. This is the species that I hoped to study without having to resort to chumming to produce feeding frenzies. I would at last find a way to study the behavior of sharks without introducing elements that would tend to skew or distort that behavior. It was time to live in their mysterious and often dangerous world—something that had never been done the way I proposed to do it. But I was convinced that to understand sharks—what their behavior meant, how they navigated, what their world looked like to them—I needed to enter their world as unobtrusively as possible, appearing to be neither predator nor (I hoped) prey.

CHAPTER 5

SWIMMING WITH HAMMERHEAD SHARKS

It was a hot summer day in the summer of 1979 in the dead-calm Gulf of California. This was toward the end of the second of five years that I was a graduate student at Scripps. This inland sea, which originates at the mouth of the Colorado River near the border between the states of California and Arizona, extends southward nearly a thousand miles before opening to the eastern Pacific Ocean near Mazatlán, Mexico. It is a slender body of water, never more than 50 miles wide, created over geological time by the Baja Peninsula, which is on the Pacific crustal plate, which is slowly moving away from the North American crustal plate. Along the western shore of the gulf is an unbroken chain of desert islands of volcanic origin that rise majestically out the beautiful blue waters of the gulf. Interspersed among the islands within this archipelago are seamounts that have yet to reach the surface through the accretion of layer upon layer of rock from volcanic eruptions. The gulf is home to a diverse fauna of fishes, sharks, dolphins, and whales, which live here in superabundance.

I was sitting in a long, narrow panga, a flat-bottomed fiberglass boat used by local fishermen, who fill them with nets to fish in these waters. I had named the boat *Pez Martillo*—Spanish for "hammer fish," the elusive species of shark I was seeking. I had spent much of the last year going from place to place looking for schools of hammerhead

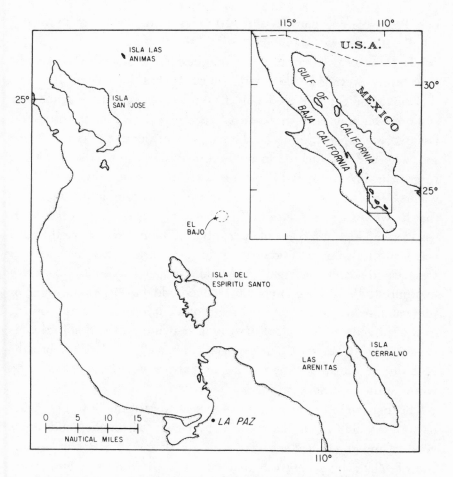

I encountered immense schools of hammerhead sharks in the Gulf of California near La Paz on the Baja Peninsula, during July and August 1979. I discovered large schools at Las Arenitas, a rocky reef on the western side of Cerralvo Island; during a cruise to El Bajo, a seamount 10 miles northeast of Espiritu Santo Island; and in the waters around Las Animas, a massive rock east of San Jose Island. (COURTESY OF THE NATIONAL MARINE FISHERIES SERVICE)

sharks. I was going to look for them at one more place. The water in front of me was perfectly flat with bright sunlight reflecting off the deep blue surface. Several hundred feet behind me was the desert island Isla Cerralvo, consisting of massive mountains littered with huge rust-colored rocks, except for a speckling here and there of green candelabra-shaped cardon cactuses. The ghostly white outline of the

Baja Peninsula was barely visible 10 miles away. We are really in the middle of nowhere, I thought.

There were three of us in the boat beside the helmsman. We were now busily putting on our masks, snorkels, and fins. We all felt sure this would be the day we would finally locate the elusive schools of scalloped hammerhead sharks that I had been searching for during the last two years. Next to me on the rear seat of the panga sat Ted Rulison, who had volunteered to help me in my research. A semi-retired surgeon in his fifties, Ted spent his spare time scuba diving in these waters. He was quite a sight in the white, tight-fitting long underwear that he wore to keep the sun off his fair skin, which was covered with freckles and highly susceptible to sunburn. The mirrorlike sea that day was a sun reflector that could toast his skin in less than an hour. He was also rightfully worried about skin cancer caused by excessive exposure to the sun. Ted may have looked odd, but he was at home in the water, easily diving on one breath of air to 60 feet before coming to the surface. In his student days, he had swum on the Stanford University swim team, and that prepared him well for "free" or breath-hold diving, the swimming technique that we would use to enter the world of the hammerhead shark.

Sitting across from us on the next seat toward the bow and facing us was Donald Nelson. Don was baby-faced, tall, slender, and also a good swimmer. He had subsidized his graduate studies at the University of Miami by selling fish that he captured while spending many days spear fishing in the Florida Keys. As mentioned earlier, he was a member of the committee overseeing my doctoral research at the Scripps Institution of Oceanography. During one of our first meetings in September 1977 after I had arrived in California, Don had shown me a couple of yellowed scraps of paper that Earl Herald, former director of the Steinhart Aquarium, had given him. Earl had died from a heart attack before he could publish his observations. The typewritten manuscript described how he had briefly seen a massive school of hammerhead sharks a great distance away from him during a scuba dive at the remote island of Las Animas in the middle of the Gulf of California.

Don suggested that I search for the hammerheads, and if I found them, to earn my degree by discovering why they formed schools. This behavior seemed odd because generally fish form schools to avoid predators—or so the theory goes—and large sharks hardly need to worry about predators. He warned me, "Take extra caution when swimming with those hammerheads because they are among the most dangerous of sharks." The graduate school at Scripps agreed about the risks involved in getting in the water with sharks. They took out a $50,000 life insurance policy for me—which I was fortunately and blissfully unaware of at the time—naming my wife as a beneficiary.

Rulison had seen the hammerheads once before at this spot and had directed us to this remote place off the shore of Isla Cerralvo. It was called Las Arenitas, Spanish for the chain of small beaches below two huge white cliffs. A couple of hundred feet from shore was a cluster of rocks parallel to the shoreline. From these rocks, the bottom was barely visible 50 feet below. It dropped off steeply to over 2,000 feet deep not far offshore.

I was the first to enter the water, with only my mask, snorkle, and fins. I sat on the side railing of the boat, lifted my mask off my forehead and put it over my face, placed the snorkel in my mouth, and fell backward into the water. The splash of my entry created a mass of white bubbles, briefly blinding my vision, but once they dispersed, I was surprised to see thirty sharks not more than twenty feet away, slowly swimming in a soldierlike formation. Each hammerhead was flanked by a companion on either side with another shark directly in front of it and one behind. The sharks were assembled into a formation that resembled a squadron of fighter planes. The members of the school were close together, only a body length apart from each other. With their flattened heads expanding to either side in the shape of a hammer, these were truly bizarre creatures.

Suddenly, my hypnotic concentration on this awesome mass of sharks was broken when I heard a loud *sphhhh* and then another *sph-hhh* sound. Don and Ted had entered the water. I looked toward the boat and saw them in the distance—I had been carried away from the boat by the current. I waved my hand back and forth to get their attention and then moved my hand toward myself to indicate that they should swim to me. They set off at once. When Ted and Don came to my side, I pointed to the sharks, which were by now barely in sight. We all stuck our heads out of the water and exclaimed, "Wow!" However, we then needed to decide what to do next. We agreed to put on scuba gear because that would make it easier to swim after the sharks. It took us no more than five minutes to get back into the water. We started out swimming in the direction the sharks had taken, moving forward in a triangular formation with a large mass of white bubbles rising from us. I swam in front while thinking to myself, "Am I a bit crazy swimming in the 'point' position of a platoon of underwater soldiers!"

We swam deeper and deeper, staying just above the rocky slope that dropped off abruptly not more than a quarter mile from the rock. The sharks were nowhere to be seen. After consuming nearly all the air in my tank, I turned around and pointed upward; we slowly began kicking to the surface. Lifting our heads out of the water, we took our regulators from our mouths and started talking excitedly about what had happened. All three of us had the same two questions. Where had they gone? Why had they left? Had they not an hour ago been swimming close to the rock? We decided to return to our land base in the city of La Paz, where we could find the materials and equipment necessary to attract the sharks to us. We would buy oily fish such as mackerels from the fishermen to make chum. The hammerhead sharks would smell the odor of the chum carried in the current and follow the scent to its source, our boat. We would also bring our portable sound-playback system. This would emit sounds made by the fish that the hammerheads ate and would summon them to us. It would take us a day to get ready, but we would return to Las Arenitas better prepared.

We returned to Las Arenitas two days later in the *Pez Martillo*, packed full with diving gear, chum, and the sound system. Just as important, we had a plan. We would tie the boat off to the same rock. Don would stay on the boat to dispense chum and operate the sound system. Ted and I would dive to 100 feet, where we would hover close to the rocks in view of the speaker while it projected the prey sounds. Here we would wait patiently for the sharks to arrive. This was going to be a scary situation to be in because we might be mistaken for the source of these sounds and the sharks might attack us rather than the other fish swimming around us. Furthermore, there was no cave or crevice on the steep rock cliff in which we could hide. All we could do would be to tread water in a vertical position, oriented back-to-back, so that a shark could not attack either of us from the blind side. Jeremiah Sullivan and I had used this tactic off the island of San Pedro Nolasco, when two immense bull sharks that were 9 feet long and weighed 400 to 500 pounds began to circle us, thinking we were not that different from sea lions and might make a desirable meal. Sea lions swooped down in the water toward the sharks at this time, barking loudly and mobbing them as birds do a predator. They appeared to be trying to drive these dangerous predators away from the waters near their shore-based colony. Luckily for the two sharks, they lost interest in us and swam away. I say "luckily" because we had reluctantly removed the safeties from our powerheads (poles containing a bullet inserted into a barrel at the end of the pole) and we were prepared to kill these beautiful sharks if they ventured any closer.

Don lowered the speaker over the side as Ted and I sat on the boat across from each other wearing our heavy scuba tanks. Once the speaker sank 100 feet down, we let ourselves roll backward into the water and began to swim downward using the cable attached to the speaker to guide us toward the bottom. Near the surface, visibility was limited to a dozen feet because of marine "snow," the name given to cottonlike aggregations of organic material that form in the ocean and are carried in the currents. It was as though we were swimming in a heavy snowstorm, except the snow did not come from above but passed by laterally, suspended in the current. The poor visibility made both of us nervous because the hammerheads might be close yet unseen. Neither of us much cared for the idea of being attacked by sharks we could see, much less by those we couldn't. Then to our amazement, as we reached a depth of 40 feet, the water became crystal-clear. Colors became blue green because most of the hues in sunlight could not penetrate to these depths through the thick layer of marine snow. We could see the speaker below us at the end of the cable. Once we reached it, we veered to the side and stopped next to a rocky cliff, from which we looked out into the void.

Close to the cliff, scattered in the water, were saucer-shaped creolefish with their sickle tails. Each was about the size of a human
hand. In shallow water, they were bright red, but, at this depth, they
were a dull gray. Even in the clearest water, the color red is absorbed
before reaching this depth. Only the blue and green light penetrates
to these depths, giving the water all around us that hue. Off the wall
about 20 feet away was a diffuse school of silver jacks, which appeared
a brownish color. The jacks, oval fish with flattened sides, darted about
in a loose aggregation. Occasionally, one would dash off to catch a jellyfish or tiny shrimp as it was carried by in the strong current. Now
the speaker began to emit a very loud and monotonous staccato drumming: *borum-bodum-bodum-borum-bodum*. Oddly, it sounded just a bit
like the theme music from *Jaws*. We turned our heads around from
side to side, expecting hammerheads to charge the speaker. I even felt
some trepidation that the hammerheads might attack us. But, as we
looked out into this blue-green world, not much happened for the
next forty minutes. Only once did we catch a glimpse of the vague
shape of a hammerhead shark swimming slowly off in the distance.
Where are the hammerheads? Why are they not attracted to these
hypnotic sounds as are the reef sharks? At the end of the dive, we
were surprised (and even frightened) as a cloud of turbid water slowly
rose out of the depths and engulfed us, reducing our visibility to
nearly zero and leaving us feeling defenseless. This water was really
cold, indicating that its source was far below the thermocline. This is
the boundary between the thermally homogeneous surface waters
that are well mixed by the winds and the undisturbed cooler waters of
the deep sea. We swam upward to avoid this mass of black water,
which stopped rising at a depth of 80 feet, made a brief decompres-
sion stop at 10 feet, and then surfaced. "No luck—no hammerheads!"
was the first thing I blurted out to Don. He grabbed our tanks and
lifted them into the boat. We then climbed aboard the *Pez Martillo*.

Don and I sat there for the longest time debating why the hammerheads were not summoned during our scuba dive. He assured me
that this was the same sound that had attracted a hundred reef sharks
at one time at Eniwetok, an island in the western Pacific where he had
studied the effectiveness of these low-frequency sounds in attracting

these sharks a decade earlier. Ted had little to say, and frankly was tired of hearing us talk on and on and not find an answer to the seemingly simple question. Why had we not seen the sharks during this dive? We had surely dumped a massive amount of chum into the water and had played the loud sounds that were supposed to be so attractive.

Finally, Ted announced he had to go to the bathroom. The only way to relieve yourself when aboard the *Pez Martillo* is to jump into the water, swim away from the boat, and go in the sea. Ted sat on the side of the boat, spun around, and then let himself into the water. He swam away from the boat along the ridge of rock, which was parallel to shore, and continued a little farther up the current from the tip of the rock. He remained in one place for a while, no more than 25 yards from us.

Don and I were still trying to decide on our next step in our frustrating quest for hammerheads, when we heard Ted shout, "They're under here!" We looked in his direction and saw him shouting over and over again, "The hammerheads . . . the hammerheads . . . they're here!" He was holding his arms up and repeatedly pointing downward with his hands to show us where the sharks were.

Our first reaction was to put on our scuba tanks. However, I hesitated. The sharks had stayed away during our dive while we baited and played back sounds. Yet now they were here, only ten minutes later. This was really odd! Then it occurred to me that we had been wearing scuba equipment before and now Ted was swimming just with a mask, snorkel, and fins. I abruptly stopped reaching for my tank, turned in Don's direction, and asked him whether he thought that the sharks might be avoiding us because we were wearing scuba tanks. The stream of bubbles exiting from our regulators produced very loud and low-pitched sounds, and the bubbles also reflected bright light while rising to the surface. This device provided air compressed by three atmospheres of pressure at our depth during the dive. The air was stored under an even greater pressure in the scuba cylinder to maximize the supply of air carried underwater for the duration of dives. However, as these tiny bubbles rose upward to either side of our masks, they became larger and larger, doubling in size every 33 feet, until they were huge when they contacted the surface. As they rose, they flattened due to the resistance of their upward movement against the water, changing the shape of the bubbles. They also moved back and forth to either side, creating a *bub-blub-bub-bulbing* sound.

Don agreed with me that the loud, low-pitched sounds produced by the bubbles would be audible and perhaps frightening to the sharks. We decided to try to follow them, not with scuba gear this time, but with only our masks and snorkels.

Don and I jumped into the water and swam toward Ted. We had to swim awhile before we reached him because the current had begun to flow more swiftly. When we did, we looked down and saw below us the vague shapes of countless hammerheads swimming in tight formation like fighter planes. In each formation, a body length behind one shark swam another shark at the same depth. Halfway between these two sharks, in line with each other, were two more on either side of the pair. These flankers were below the others. Together, the four sharks made up a diamond-shaped formation. The school's many formations, all separated by the roughly equal distances and all swimming in the same direction—usually against the current—presented a stunning spectacle of organization. Every now and again, a shark would roll on its side and became more visible as light flashed off its white belly. I hypothesized that they might be doing this to better see us through the marine snow in the water that impeded their vision as much as it did ours.

I took several deep breaths through my snorkel, stretched my arms out in front of me, and propelled myself downward, kicking vigorously with my fins. These were no ordinary fins, but designed especially for "free divers" unburdened by heavy scuba tanks. My fins were lightweight, long, and flexible—less than 8 inches wide but close to 5 feet long, almost as long as I am tall. The fins did not stay rigid when I kicked, but bent into a sinusoidal wave, such as the way a snake moves its body. This type of swimming is termed anguilliform, after the order of eels, which use this method of locomotion.

Down, down, down I swam to the top of the school at a depth of 30 feet. My heartbeat quickened, but I didn't stop. I continued downward, passing between the sharks at the top of the school and moving into the center of the school 60 or 70 feet down. Below me were sharks all the way to the bottom. Looking to the side, there were sharks as far as I could see. More sharks than I could possibly count. Maybe fifty sharks, maybe a hundred, perhaps even more.

Reaching the center of the school, I stopped kicking and let myself float to an upright position and then rotated my body 360 degrees, looking at the expanse of the school. This was probably 70 feet below the surface, which was not even visible to me now. There was an eerie silence in the center of the school. The sharks were so close to me I could have reached out and touched the one above me, the ones on each side, and the one below me. I dared not since hammerheads were labeled "man-eaters" in every shark book that I had ever read. In fact, David Baldridge, who compiled a file of attacks worldwide for the United States Navy, rated the hammerhead the third-most-dangerous shark species in the world, just below the white and tiger sharks.

Yet these scalloped hammerheads didn't appear terrifying at all, but elegant and graceful. The sharks swimming above me were particularly beautiful. Their bodies had a bronze gloss when viewed from below and illuminated from above. They swam so effortlessly, moving their tails slowly and sinuously, all the while staying in a perfect formation with each no more than one body length from the other. Their spacing ratio was comparable to that of a school of sardines, given the spacing of their separation relative to their body length, but because they were so large they seemed more separated—more like a school of miniature submarines!

The hammerheads passed by me, one by one, swimming ever so slowly. At times, they moved their eyes, bulging outward from the end of their laterally elongated heads, so that they could better see me. I stared at the jaws of the biggest shark above me, perhaps 8 feet long, and was impressed not with how large its mouth was, but how small. It was restricted to the center of its head and did not extend all of the way to the edges of the hammer-shaped rostrum. These were not jaws designed for consuming something the size of humans, but were just large enough to swallow the small fishes and squids that live in the surrounding ocean.

I resumed kicking, propelling myself slowly toward the surface. As I passed by sharks on either side of me, the ones above me veered slightly to one side to avoid swimming into me. Reaching the surface, I cleared my snorkel, forcing the water out of it with a puff of air, and then breathed deeply several times. Relaxing into a normal breathing pattern, I reflected upon what a golden opportunity this was for me to enter the underwater world of sharks.

As soon as Don saw me swimming at the surface, he made his dive into the school of sharks. We had decided to alternate making free dives. This was a technique full of risk, and one that should only be used by powerful swimmers. One of us always watched the other from the surface. We were aware of the danger of blacking out and feared it could happen to us here because we were diving very deep to enter the schools. If one of us were to become limp and stationary in the water, the other would dive down to fetch him, drag him to the surface, and revive him with mouth-to-mouth resuscitation. Thankfully, such an emergency never occurred to us when we were among the hammerhead sharks, though at times we descended over 100 feet to attach tiny tracking devices on the sharks, using a pole spear.

We made dives into the school of hammerheads all afternoon. I recorded observations while at the surface between dives using a small pencil attached to several white tablets held together by a binder ring. This was tucked into my wet suit while diving. I needed to record the size of the sharks, their sex, and to describe what they were doing in these schools. Always on my mind was the question "Why were they in these schools?" Surely, it wasn't for mutual defense, since there was little possibility that an adult shark could be eaten by any conceivable predator. Most ichthyologists explain schooling as a way to avoid being eaten. There are more eyes and ears in a school of prey to see or hear a predator, too many moving targets for the predator to focus on a single individual. Furthermore, once entering a school, that predator might be distracted from picking out an individual to capture by the nearness of the many other members of the school, darting collectively here and there in confusing motion.

By the end of the day, we were both exhausted and cold. You always lose body heat when spending four or five hours in the water—even in the Gulf of California during late summer when the water temperature at the surface can be close to 90°F and feels initially like a warm bath. You quickly lose heat when diving to depths of 60 to 100 feet, below the thermocline where the water temperature is in the mid 70s to low 80s. I often shivered momentarily as I swam downward through the thermocline and entered into waters that cooled with increasing depth.

That evening back in La Paz, we met Flip Nicklin, a photographer from the National Geographic Society, who had just flown in from San Diego. He and his father operated a scuba shop in La Jolla California. Flip wanted to take pictures of hammerheads for an article on sharks in an upcoming issue of *National Geographic*. Although a top-notch scuba diver, he was also an avid free diver and had competed in the world championship of spear fishing. Divers from all over the world would come to these competitions; the winners were those who caught the most fish each day. Flip had done very well in the competition when it was held in Southern California.

Don, Ted, and I had dinner that night at Los Arcos, a lovely Spanish Colonial hotel in downtown La Paz with a restaurant looking out upon the Malecón, the avenue along the waterfront with its white, sandy beaches littered with the small pangas of local fishermen. Flip had chartered a boat for a ten-day cruise from a newly built marine laboratory in La Paz, Centro Interdisciplinarios de Ciencias Marinas, CICIMAR for short. This institution was located on the outskirts of town on the shore of La Paz Bay and consisted of several buildings arranged in a square to surround a quadrangle. It was run by the Instituto Polytécnico Nacional, part of the government university system. At dinner Ted told us about another location where he had seen the sharks. One that was a seamount that rose steeply from the deep water to within 60 feet of the surface. Called El Bajo, meaning "the shallows," it was 10 miles east of a pair of massive rocks called Los Islotes off the northern end of Espíritu Santo, an island near the city of La Paz, and itself fully 7 miles distant from shore. Ted said that this was the best place to find hammerhead sharks in the Gulf of California. We decided that we should visit this site to study the hammerheads.

When we arrived at the boat, we were impressed by its raffish appearance. It looked more like Humphrey Bogart's *African Queen* than a modern research vessel. It was as wide as a barge, with a short mast in front and a large canopy over the rear that offered some protection from the sun. The *Juan de Dios Bátiz*, (named for the founder of the university), was made of ferrocement, iron screen impregnated with concrete. Here and there on the boat, the concrete had cracked and the iron had been exposed to salt water, which had rusted the metal and left an orange-brown residue on the side of the boat. There were two cabins, one for the scientific party and another for the crew, but nobody used the bunks inside these cabins because the air inside was too hot for sleeping, even at night. Everyone slept on the deck, lying on a blanket or in a sleeping bag.

On one side of the deck was the galley, consisting of a gas stove fastened to the side of the rear cabin. There was a tiny head (bathroom), which was hardly ever used because it rarely worked. The crew went to the bathroom over the stern when nobody was looking. These would be our rudimentary living quarters for the next ten days when not swimming with the sharks.

One evening later, Jack McKinney, a well-known underwater photographer, joined us aboard the boat for dinner, invited by Howard Hall, who was making a nature documentary for the *Wild Kingdom* television show. Years later, I heard Jack describe that evening in his opening monologue while hosting the San Diego Film Festival. He said that the first thing he saw upon boarding the boat was a rusty old stove with a big pot on top containing a dead chicken with its two legs sticking out of the pot. He was going to have to eat that! He walked along the side of the cabin and noticed that part of it was charred, having burned in a recent fire. There was bedding all over the deck, scuba gear here and there, and people everywhere. The place was filthy and, he declared, in fact smelled. He couldn't understand how Klimley and Hall could live in conditions so crude that rats wouldn't tolerate them. The conditions really weren't that bad.

We departed from La Paz early on the following day, but did not arrive at the seamount, close to 40 miles away, until near sunset. It took some time to find the spot. That was before the days of the hand-held GPS, or Global Positioning System, and we had to use ranges that Ted had given us to find the seamount. We drove the boat in a straight line, keeping a mountain on the mainland, barely visible 25 miles away, behind the rocks on the north end of Espíritu Santo Island, which were 15 miles away. We proceeded along this course until a prominent peak on the mainland west of us was directly behind the gap between Los Islotes.

Three of us drove the panga in front of the *Juan de Dios Bátiz*, looking for the seamount with a depth finder. To our astonishment, a black mark appeared on the fathometer and the line moved upward steeply, indicating a mountain rising precipitously from nearly a mile down to a depth of 55 feet. Then we could just make out the highest peak of the underwater ridge in the clear blue below. All around us, the flat surface of the water was being disturbed by the splashing of hundreds of small fish feeding on jellyfish and other planktonic animals concentrated near the surface around the seamount.

Suddenly, there was a loud *vrooooom* as a predator dashed into these aggregations of fish and individuals scattered to all sides, trying to avoid being eaten. It was probably a dolphinfish, or *dorado* in Spanish. This predatory fish is brilliant golden green and possesses a huge dorsal fin that extends down the back, and is not to be confused with the mammalian dolphin, which is an air-breathing toothed whale. Then a male dolphinfish with its bulbous head shot across the surface, partially visible as its laterally streamlined body cut through the water chasing a flying fish. It had launched itself into the air and was gliding close to the sea surface, touching down momentarily and then propelling itself again into the air by rapidly vibrating the lower lobe of its tail in the surface water. This place was packed with fish. However, it was getting dark quickly, and we would have to wait until the following morning to look for hammerheads.

The water was absolutely flat when dawn broke. We separated into two groups, each consisting of three divers and the driver of the panga. Don, Flip, and I were in one group. Ted Rulison, Jeremiah Sullivan, and Felipe Galvan-Magaña, a precocious Mexican graduate student very interested in learning more about sharks, made up the second group. Our first idea was to swim over the seamount, but once we made eye contact with the rocky ridge, we had to kick furiously to maintain our place and eventually were swept away from the seamount. The currents at the seamount were often twice as fast as a world record-holder could swim. It was hopeless to try to swim against such currents. We knew we passed the seamount because the *clickity-click* sounds of the shrimp that inhabited the rocks faded as the bottom fell away below us.

Our next plan was to drive in the direction from which the current was coming, watching the fathometer to see when the bottom began to drop, and then getting in the water and letting the current carry us over the seamount. We would emulate the plankton that were being transported continuously past the seamount in the current. Three of us would stay in a line perpendicular to the direction of the current, just within sight of each other, and would alternate diving down and looking for the hammerheads. The driver had to be careful to keep us constantly in view. Each diver faced the risk that he could be swept away and lost. This was a much more immediate danger than being attacked by sharks.

Once we entered the water up-current of the seamount, we would promptly look around for hammerheads. As far as we could see, there were loose schools of small fishes dashing here and there to swallow the many little jellyfish, shrimp, or larvaceans born in the water. The latter secrete a tiny net of mucus filaments, through which they draw water by beating their tails and capture miniature species. Throughout, we could see yellow-finned pompano, greenish-silver jacks, and bullet-shaped silvery mackerel. As we drifted toward the seamount, we suddenly saw a tightly packed mass of large snappers. These predatory fishes form schools during the day and hunt individually at night.

Manta rays, twenty feet across from wing tip to wing tip, would occasionally break the surface in front of our bow. As we approached, they would roll over to complete a reverse somersault in slow motion, exposing their highly reflective undersides. Either they wanted to get a better view of us or to frighten us with the bright light reflected off their bellies. On some of them, we could see one or two remoras, long gray fish with a toilet plunger–like sucker on the underside, attached to the manta's skin. We drifted by a school of small bonito tunas that was 100 yards long and 25 yards wide. If we got too close to the school, a blacktip shark would dash up to one of us, hunch its back, and start moving its jaws up and down as an aggressive gesture: it reminded me of a shepherd protecting his herd. The sharks seemed protective of this source of food.

During this trip, we even saw a blue-spotted whale shark. Unlike an air-breathing whale, which is a mammal, this shark, the largest fish in the oceans, extracts oxygen out of the water with its gills. The only similarities to a whale are its size and its propensity to consume zoo-plankton. However, unlike baleen whales, it often feeds on small fishes as well. Ted swam down to the whale shark. His length with long flip-pers must have been 8 feet, and the shark was five times that length: close to 40 feet long.

Then a huge school of hammerhead sharks came into sight. There were sharks as far as we could see, so numerous that it would be impossible to count them. Even though the water was clear, the ham-merheads often blended into the background, thanks to their gray col-oration, shading from dark gray on top to light gray on the bottom. The dark gray appeared lighter when illuminated from above, and this color perfectly matched the lighter gray underside.

Although this uniform gray color would make it difficult for prey species to see this predator, it would be a disadvantage to the shark when it needed to be seen by members of its own species. A member of the school would occasionally accelerate out of its anonymous posi-tion and perform a complicated "gymnastic" movement, which pro-duced a unique series of pulses of light reflecting off the shark's body. This always attracted my attention and surely that of other sharks. These sharks were not just swimming in formation, they were inter-acting, perhaps even communicating with each other in a strange but intelligent way. The sharks often accelerated upward or downward while shaking their heads and the front parts of their bodies spasmodi-cally. The shaking could be continuous, discontinuous, to one side, or to both sides.

The most acrobatic shark behavior of all resembled the backward somersault and full twist performed by a human gymnast. I call this behavior the *Corkscrew*. This involved the shark's accelerating explo-sively into a small circular path less than a body length in diameter

Corkscrew

In the *Corkscrew*, a female hammerhead shark accelerates into a looping trajectory and rotates her body completely. The *Corkscrew* is a threat display that functions to drive smaller subordinate females away from the center of the group, allowing the larger and more dominant females to be noticed by a courting male. (COURTESY OF BLACKWELL PUBLISHING)

while rapidly twisting its body in one complete rotation. The shark completed this gymnastic maneuver in less than a couple of seconds. Almost as spectacular was the *Torso Thrust*, in which the shark propelled itself forward in a jerky manner with exaggerated beats of the tail while thrusting its midsection and rotating a clasper to one side. Each of these three behaviors resulting in pulses of light reflecting off the shark's belly that were visible at a great distance. I believe that they are signals aimed at others within the group.

I wanted to know the sex of the sharks in the schools. This information might provide insight into the function of the schools. Might they be leks, groups of males that fight each other for access to females? This social system was common among birds and some mammals. To determine this required that I make especially deep dives, swimming down through the school, righting myself, and looking upward at the underside of each member. (As explained earlier, a male shark differs from a female in that he possesses two claspers, appendages on the underside that are inserted into the female vent or cloaca during mating.) Almost all of the hammerhead sharks in the school were females. Seen at an arm's length, many of the females had white contusions on the top of their heads and the forward parts of their bodies where their denticles, the toothlike scales covering their body, had been scraped off. On the underside of their jaws, I noticed dark patches of the same shape and assumed that these were similar in nature but had become colored when healing.

The cause of these scars became apparent when I saw the sudden movement of a shark downward, hitting its snout against the back of the shark immediately below it. Later, I noticed other sharks ending their *Corkscrew* by contacting a nearby shark on its back. These were female sharks fighting among each other. But why were they doing this? Could they be competing for a central position within the school in a fashion comparable to the way male birds such as grouse compete for space within their displaying grounds? Could the males be penetrating the schools, advertising their sexual prowess with *Torso Thrust* and choosing the biggest females in the center of the schools? To find out the answers to these questions, I would need to develop some way of measuring the sizes of the sharks. I also needed a video system so that I could document the behavior of the sharks and then analyze it over and over again in the laboratory.

Don and I talked during lunch the first day at the seamount about how to estimate the size of the population of sharks at the seamount. Lunch always was an obstacle to work on the *Juan de Dios Bátiz*. We wanted to eat quickly and get back to work, while the cook always wanted to prepare a large *comida*, as was customary in Mexico. In preparation for the cruise, my wife had pickled a few dozen hard-boiled eggs for me and packed them in four 1-gallon containers. My supply would enable each of us to have two eggs per lunch for the duration of the cruise. Unfortunately, Don, Ted, and Flip hated my nutritious pickled eggs, instead favoring peanut butter and jelly tortillas. While eating our chosen foods, we decided how to estimate the size of the population by marking the sharks with color-coded tags.

For years, ecologists have used a method called the Lincoln Index to estimate the size of populations of animals. You might use this statistical sampling technique to estimate the number of marbles on a table. Imagine that there are one hundred glass marbles in front of you. Take ten and put them to the side, spray them with red paint, and let the coat dry. Now mix these colored marbles with the unpainted, transparent marbles. Now close your eyes and pick up ten of the marbles, each from a different place on the table and put them to the side. Open your eyes, and look at the marbles on the edge of the table. The odds are you will have one red marble and nine clear marbles. This is because your ten marked marbles comprise one tenth of all of the marbles. Thereafter, even when the number of the marbles on the table changes, you can estimate how many marbles are in front of you by simply multiplying the number of marbles spray-painted and mixed with unpainted marbles in the first place (ten in this case) by the fraction, consisting of the number of marbles that you take in your hand (ten in this case) divided by the number of these marbles that are painted (one). Say you paint the same number of marbles and mix them with an unknown number of marbles on the table. If you pick up ten marbles, and find that two of the ten are painted, there are fifty marbles on the table.

We intended to estimate the number of sharks using the same technique. We couldn't spray-paint the hammerheads but could mark them in another way. We could dive into the middle of the school, pick out a shark, and attach a color-coded streamer or "spaghetti" tag to it. We would have to carry along a pole spear with a small slot in its tip, in which fit a small stainless steel barb attached to monofilament line threaded with short sections of vinyl tubing of different colors. The barb would be inserted into the thick back musculature of the shark. An individual could then be recognized by the unique color code on its tag. Instead of grabbing marbles in one's hand, we drifted by the seamount and counted the tagged and untagged sharks that we passed. The color-coding was necessary so that we did not count the same marked shark twice.

Of course, the first task was to attach these colorful spaghetti tags on the sharks. That night I assembled thirty tags. On the following morning, Don and I mounted a tag on each of our pole spears. We carried the rest of the tags with their metal barbs inserted in slits on rubber wristbands: the ring of colored tags streaming down our arms had the appearance of jewelry. We jumped off the panga, well up-current of the seamount, and began to drift with the current. Before long, the massive school of hammerhead sharks came into view. I made the first free-dive into the school while Don watched. I was apprehensive now because I was no longer simply observing the sharks but was inserting a sharp metallic barb into the shark's back. This might hurt. Suppose the shark reacted by turning around and trying to bite me.

By the time I swam down to the depth of the school, it had passed by. I came to a standstill underwater, looking at the sharks swimming away in the distance, and swore to myself in frustration while biting down on my snorkel. It took over half a minute to reach the surface. Here I took several gasping breaths of air. I had been underwater too long and needed oxygen badly. Don expressed his sympathy; he had seen me narrowly miss the school.

I next kicked into a position where the sharks at the front of the school, 80 feet down, were coming slowly in my direction and dove down. By the time I reached 90 feet, there were sharks all around me. I began swimming slowly on the horizontal plane, like a member of the school, moving at the same speed and in the same direction. I picked out a large individual 6 feet away and kicked a little harder to catch up with it. It moved out of reach, however, so I had to swim downward again to the next shark in the school. I now grabbed the spear with my left hand and pulled the rubber sling forward with my right hand until it was taut. Gripping the shaft of the spear with my right hand, I then extended my arm forward, aiming the spear downward so that the tag would go onto the shark's back, just behind and to the side of its top fin. When the tip of the spear was less than a foot from the top of the shark, I relaxed my grip so that the rubber sling recoiled, propelling the spear through my loose handgrip toward the shark's back.

Thooom was the sound of the spear contacting the shark. It suddenly accelerated forward in response to its being tagged by me. The shark darted out of the school but slowed down as it made a large circle and, by the time I reached the surface, was again swimming within the school, oblivious to the tag on its back.

The blue, orange, yellow, white sequence of colors on the tag could easily be seen from the surface. I now reached for the next of the dozen tags that I carried on the rubber band on my left wrist. Don swam over to congratulate me. He said I had seemed to stay down forever on my last dive. And that was the way it was when tagging hammerhead sharks at the seamount. You could stay down for a long time when tagging sharks because you had a single, pressing objective: get the tag off your spear and onto a shark.

Don and I alternated diving into the schools and tagging sharks. By the end of the morning, we had placed tags on 21 hammerhead sharks. That afternoon, we observed only 9 of the tagged sharks while counting 225 sharks during our dives. We divided the number of sharks observed at this time (225) by the number of individuals with spaghetti tags (9) and multiplied this number (25) by the number of tags that we had attached to sharks during the morning (21) to arrive at an estimate of 525 sharks present on the north side of the seamount that day. This was a lot of sharks concentrated above an acre of the northern slope of the seamount where the water ranged from 80 to 110 feet deep—far more than anyone had ever seen underwater in one day.

Oddly enough, I felt little fear of being attacked while swimming among schools of these reportedly dangerous hammerhead sharks. Indeed, the public was still deathly afraid of hammerhead sharks. Soon after returning from this expedition, the News Service at Scripps sent me a newspaper clipping that reported a massive stampede of people leaving the water off South Beach in Miami when informed by a life-guard that a single hammerhead shark had been spotted swimming close to shore. While swimming within the schools later that day, I wondered why humans were so afraid of sharks. These hammerheads surely weren't going to eat people. Most of the school members were no larger than an adult human, and their mouths were no more than 9 inches across. The title of "man-eater" is really an inappropriate name for the scalloped hammerhead, as well as for the majority of sharks in the world. I was just beginning to learn about hammerheads. Their set was the schooling behavior that most aroused my curiosity. Yet there is a motivation for even smaller species to attack a human—to defend their personal space. When swimming alone, a scalloped hammerhead will attack when a person rushes toward it and it is deprived of space to escape. When frightened, any shark (like many other animals) will exercise one of two behavioral options—fight or flight. On the other hand, the same individual would be very docile when swimming in the company of its schoolmates at the seamount. Of course, there are different species of hammerheads, nine in fact, ranging in size from the yard-long bonnethead to the great hammerhead, which often reaches close to 20 feet in length. The scalloped hammerhead, the most common species, is usually the size of an adult human being and similar in size to the majority of the members of the genus. The great hammerhead, easily identified by its prominent sickle-shaped dorsal fin, is usually observed swimming alone, and thus might be more likely to attack humans than a schooling scalloped hammerhead shark. The great hammerhead's attack is likely defensively motivated because it has a relatively small mouth opening for its size. The width of jaws of a 13-foot-10-inch individual that I captured measured only a dozen inches, hardly wide enough to swallow a human.

I was just beginning to learn about hammerheads.

CHAPTER 6

SOLVING THE MYSTERY OF HAMMERHEAD SCHOOLS

It was on only the second day of our trip to El Bajo Espíritu Santo during the summer of 1979 that I began to wonder why the hammerheads joined in these great gatherings, and why they assembled at this seamount and not elsewhere along the coastline or in the vast expanse of deep ocean. Questions such as these came naturally to me as I drifted across the seamount marveling at the bounty of life that made it home. Answering this question was to become the objective of my doctoral research studies.

Seamounts are common throughout the oceans. Most of these underwater mountains are formed over thousands of years by molten basalt extruded through the earth's crust during volcanic eruptions that have yet to break the surface and become a full-fledged island. Chains of these are formed parallel to the direction of movement of the massive oceanic plates as they pass over a location where molten basalt leaks upward from the earth's mantle underneath the seafloor crust. Seamounts rise out of the sea to become larger and larger islands as they grow from accretion of basalt to form chains of islands of islands (archipelagoes)—the chain of Hawaiian islands is a prime example. Seamounts serve as gathering points for fishes throughout the oceans.

After vigorously propelling myself downward 70 feet early that day, I stopped kicking and let my body rise to an upright position and looked around, peering far out into the clear blue water surrounding me. Behind me and just visible was the massive rounded gray pinnacle of the Espíritu Santo seamount. Along the edge was a diffuse school of small oval red creolefish, which periodically dashed upward and together when a white cloudy mass filled the water and began to drift toward me. Many males chased each female upward, releasing their sperm as she released her eggs in order to fertilize them. This mass of animal gametes made the water so cloudy and opaque that it was impossible for a while to see anything clearly that was more than a few feet away. Before me in the distance, I could barely make out the vague shapes of one or two scalloped hammerhead sharks. As I drifted out of sight of the seamount, the water became clearer and it was possible to look far into the distance. I now saw a huge ordered mass of hammerhead sharks moving slowly toward me. The sharks first swam in front of me, but then past my side, and finally around and behind me. Now they were everywhere. There were far too many hammerheads to count during my minute and a half on one breath of air.

I came up for air and floated in the warm water, wondering what could possibly explain why hammerhead sharks formed these schools. Good reasons existed for skipjack tunas, smaller than a football, to pack together in the huge school, the length of a football field, that was always present at the seamount during the day. There were many predators—blacktip sharks, sailfish, and marlin—at the seamount. Two or three blacktips often patrolled the edges of the school, eager to capture any tuna that strayed outside. Sailfish and marlin were also present over the seamount, staying put by slowly moving their pectoral fins. Billfish were formidable predators that could dash through the schools, swinging their long rostrums, swordlike extensions of their heads that resembled medieval broadswords, back and forth, striking and killing school members. However, I could think of no predators of the adult hammerhead sharks that inhabited the seamount for any length of time. Only once, many years ago, had fishermen witnessed a great white shark large enough to eat a hammerhead slowly swimming at the surface above the seamount.

Scientists had frequently studied the structure of these apparently defensive schools by observing certain fish while in captivity. Long, slender, silverfish, topsmelt, anchovies, and herring, always lived inside schools. Schools of these diminutive fish could be kept in a laboratory aquarium equipped with lights and video equipment. The lights were positioned above the tank at a known distance from the bottom with a video camera that recorded the size and position of shadows cast by the schooling fish. From this information were determined the sizes of individuals, their swimming directions, and the distances to their nearest neighbors. The researchers found that members of these defensive schools were usually similar in size, stayed close to each other, and swam synchronously in a common direction. Such fish moved together like a phalanx of Roman soldiers marching into battle.

The scalloped hammerhead sharks in the schools that I now looked at were much larger than the skipjack tuna. Don Nelson and I were both slightly over 6 feet tall. While at the surface, one of us would look down at the other, swimming far below and next to a member of the school, and estimate the shark's size using the other as a human yardstick. (Of course, our heights were in feet so we had to later convert our estimates to scientifically acceptable meters.) We found that the sharks were often as long as we were tall, or longer. Yet our impression was that there were also sharks smaller than us in the same school. What also interested us was that, in contrast to the tunas, the sharks did not all act alike, but sporadically performed individual acrobatic displays such as the *Corkscrew* described earlier, a reverse somersault and full twist in platform-diving terminology. However, these sharks acted so quickly that it was hard to estimate their size and exactly where they were situated in the school. These groups of sharks did not seem at all like the defensive schools of smaller fishes that other scientists had studied. Although the hammerheads might be as long as Roman soldiers were tall, the sharks were not all acting in concert in a way comparable to soldiers in a phalanx marching into battle.

If these schools were not protective, what could their function be? One possibility that had occurred to us was that they were mating aggregations. If that were so, we reasoned that all of the members of the school should be capable of mating. But how did you ascertain this? First, you needed to know the length at which males and females became reproductively mature. I could measure the lengths of sharks caught by local fishermen and establish the state of the reproductive organs of these sharks. Reproductive readiness could then be equated with a threshold size. But then you would also need to know the sizes of the sharks swimming within the school, and taking those measurements would not be easy, as I soon found out.

Our use of each other as a yardstick provided only a rough estimate of the size of the sharks below. The resolution of our size estimates was too coarse. We could conclude that the shark was similar in length to the swimmer or perhaps three-quarters of his length, but not seven- or nine-tenths his length. We needed a way to make more accurate measurements.

The resulting measurements should not require conversion from English to metric units. Scientists liked the metric scale because all of the units are related to each other by tens, that is, there were 100 centimeters (ten times ten) in a meter and 1,000 meters in a kilometer. This always makes multiplication and division easier. Units in the English scale of distance are not related to each other by multiples of ten. For example, there are 12 inches to the foot and 3 feet to the yard.

Don and I agreed that a good plan was to convert a pole spear to a measuring stick. We could swim down to a shark with the spear in hand, hold it along the shark's side, and judge its length using the spear as a distance reference. I wound black electrical tape around my yellow pole spear, alternating each 10-centimeter-long hand-width of black tape with a similar length of the uncovered yellow spear. The spear ended up looking like a thin barber's pole 2 meters long.

It was now time to try to measure a hammerhead shark. I took an extra large breath of air, swam down 60 feet to the top of a school, and began swimming alongside a small female shark. I held the front of the spear even with her head and then looked back on the rest of the spear for that band of black or yellow stripe even with the her tail. After noting the placement of the band on the spear, I looked back at her head and, to my surprise, she had advanced far beyond the front of the spear. My measurement of this shark's length would be wrong, an underestimate. This was because the shark had moved away from the end of my spear while I looked backward for the position of the tail in relation to the spear. This technique wasn't going to work on moving sharks.

After returning from Mexico, I asked colleagues at Scripps if anyone knew of a better method of measuring the dimensions of distant objects. Down the hall from me in the Marine Biology Building was a scientist who was recording the densities, or numbers present in defined areas, of animals in the deep sea. I told him I needed to find a way to measure the hammerheads while they were swimming in a school. He said that the pioneering biologists, who first descended into the deep sea within submersibles, shared my problem. They wanted to have a fairly accurate idea of the size of the organisms scattered on the seabed that they could see through their portholes. These scientists could not exit the submarine to collect samples in the cold and pressurized environment of the deep sea. These deep-sea biologists seized upon a technique called stereophotography and used it to determine the size of these objects and the distances to them.

Stereophotography works in the following way. Two photos are simultaneously taken of an object with an apparatus on which are mounted two identical cameras. These cameras are separated a fixed, precise distance from each other. This fixed distance will become a reference, or yardstick, for measuring any distant object. You must simultaneously take two pictures of the object, which must be in full view, oriented at a right angle to the camera. Next, you need to examine the two photos. There will be a slight separation between the same point on any object in the two images, which is equivalent to the separation between the two cameras. We measured the slight separation between the snout of the swimming shark on the two photos. This tiny distance on the photo is equivalent to the large distance between the cameras. One now must measure the distance, from snout to tail, on the image of the shark on a single photo. One then finds the number of times that this tiny separation distance, which is equivalent to the large separation between the cameras, fits into the shark's image on the photo. The length of the shark is this multiple times the distance separating the two cameras. It sounds more complicated than it is. You will understand this technique better after I describe how I used my stereocamera to measure the sizes of the hammerhead sharks swimming in the schools.

FRONT VIEW

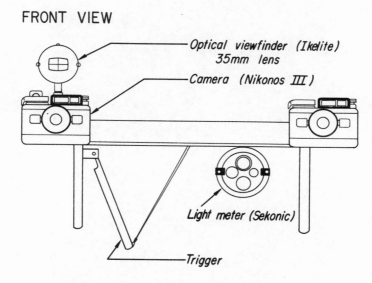

I measured the lengths of the free-swimming sharks with a stereocamera, made from two submersible cameras fastened to the base of an aluminum angle that I held by two handles. I centered a shark in the optical viewfinder and squeezed the trigger, simultaneously snapping two photographs of the shark. The image of the shark on the photograph taken by the second camera was always slightly behind the image of the same shark on the first photo. The space between the two images was equivalent to the separation between the cameras and could be used as a reference to measure the length of the shark. (COURTESY OF SPRINGER-VERLAG)

I quickly discovered that there were no stereocameras commercially available for me to use in the ocean. Furthermore, the cameras, a critical part of the apparatus, were too costly for me to purchase. However, the National Geographic Society lent two Nikonos underwater cameras to me with a grant from them to study the hammerhead sharks. I spaced the two cameras half a meter (20 inches) apart on both ends of a small section of aluminum angle. The cameras were bolted to the angle using their tripod mounts. I purchased a viewfinder, or scope, and mounted it on top of one of the cameras to ensure that the image of the hammerhead shark would be centered in the photograph taken

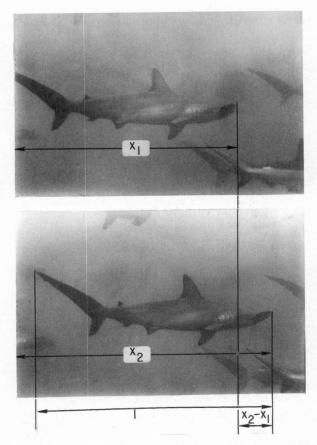

Two photographs of the same female hammerhead taken by the stereocamera. Notice that the shark's snout in the top photo is closer to the left side of the image than in the bottom photo. In this example, the subtraction of the top measurement (X_1) from the bottom measurement (X_2) is equal to one half-meter. You can determine how large the shark is by calculating how many times this reference distance (X_2-X_1) fits into the length of the image of the shark in the photo. (PHOTO BY AUTHOR)

by that camera. The two cameras were fired simultaneously by plastic-coated, stainless-steel cables passing through nylon-lined ferrules and attached to a trigger. This in turn was attached to one of two handles extending downward from the aluminum angle. I got the plastic, ring-like ferrules from an old fishing rod. The cables were connected to the camera triggers. I visited the local bicycle shop and purchased two bicycle turnbuckles, which I used to adjust the relative lengths of the cables so that the cameras could be triggered simultaneously.

I brought the stereocamera on my next expedition to the Gulf of California during the summer of 1980. Joining me that summer was Steve Brown, an undergraduate at the University of California, a shy and gifted musician, who was bright, capable, and keenly interested in the behavior of sharks. He now works for the Florida Department of Natural Resources overseeing the shark fisheries in Florida. Also with me that year was sixteen-year-old Scott Michaels, who at this young age already had a virtual encyclopedia of shark knowledge, despite having grown up in Nebraska far away from sharks and the ocean. It has so often been the case in my experience: the allure of sharks is greatest for those living in the plains or mountains farthest from them. I have heard from people living in Colorado, Czechoslovakia, the interior of Canada, Germany, and Switzerland who want an opportunity to learn about sharks.

This year, we set up a camp on a tiny island, Isla Pardito, which was only 12 miles away from the seamount. This was a remote place nearly halfway between the cities of La Paz and Loreto, over 150 miles apart from each along the coast of the Baja Peninsula. The mainland, 7 miles away, was composed of steep, impassable cliffs leading to mountains, which were surrounded by sweltering and parched deserts. This tiny island, the size of a football field, was nestled in a channel between two larger islands, Isla Francisquito and Isla San José. They conveniently blocked the strong winds that sometimes agitated the sea. They made the surrounding waters safe for the fishermen, *pescadores*, to fish for sharks. These men, with faces wrinkled from the constant exposure to the sun, would venture out into the surrounding waters each day in their long pangas filled with large plastic monofilament nets. These were used to capture a variety of sharks, ranging from little gray smoothhounds or horn sharks with their spiked fins to larger blacktips, silkies, and hammerheads, to an occasional monster tiger or white shark.

This little island was home to the families of three sons of a fugitive from the Mexican civil war. They were friendly and gracious people, partly the result of living in great isolation, for the nearest town was far away across the water. The sons were large men, tanned from spending so much time fishing in the sun and just as interested in the behavior of fishes as we were. They had four or five boats, which were hauled up on a small white crescent-shaped beach when not at sea. Just behind the beach was a long storage shed with a thatched roof used to store salted shark meat and fishing gear. A small square building made of concrete blocks served as a communal dining hall. It was situated above the beach on the slope of the hill that comprised the island. Scattered higher among the candelabra-like cardon cactus on the slope were three or four small houses in which the forty-five inhabitants of the island slept. At the top of the hill was a small chapel with a white cross on its roof. Once a week, the island was visited by a man from La Paz with a panga, on which he loaded the salted fillets from the sharks captured that week. In return, this trader left several large plastic jugs of water, for there was no natural source of fresh water on the island, and an occasional 55-gallon barrel of gasoline, needed by the fishermen to run their outboard engines.

Our camp was situated to the side of the beach at one end of the island. Here we set up a large square tent. There was ample room to spread out our sleeping bags on one side of the tent and have on the opposite side two collapsible benches, on which we kept our underwater video housing and monitor for examining our video records of the schooling sharks. Outside of the tent was workspace of similar size covered by a square tarpaulin, suspended between four poles. Two collapsible card tables were placed here. On one was equipment for developing film and a microscope for viewing the stereophotos of sharks; on the other was our stove for cooking meals. In camp, we had great privacy, except that our sleep at night was often interrupted as family members walked by on their way around the island to a tide-swept rocky shore, which they used as a bathroom.

As guests of the three families, we spent most of the summer visiting the seamount each day in our panga. During these trips, we took stereophotographs to determine the sizes of the hammerhead sharks within the schools and recorded video to describe the behavior of the sharks in the schools. When the seas were too rough to visit the seamount, we stayed at camp and analyzed our samples. The fishermen came back to camp with their catch late in the day. We timed our return so that I could examine the reproductive organs and measure the sizes of the hammerheads caught that day.

As soon as the sun rose, illuminating the beach and the surrounding waters, we pushed our panga, loaded with dive gear, stereocamera, and video recorder, off the beach and onto the water. The helmsman pulled the cord, starting the engine, and drove us around the island, out the channel between the two islands, and offshore toward the seamount. When we arrived early in the morning, the water was as flat as a mirror, except for an occasional knifelike parting of the surface by a dolphinfish rushing after a flying fish bouncing across the surface. Nobody was to be seen on the horizon. Either Steve or Scott would jump in the water with the stereocamera. I would get in the water with the video housing. The person remaining in the boat would begin slowly rowing to keep the boat close to the two of us. Below were the hammerheads. Often they swam so close to the surface that you could reach down and touch them. It was possible then to take stereophotos and record video near the surface. The sharks often rotated their widely separated eyes upward to examine me as they passed. They were observing me, just as I was observing them.

As summer passed into fall and the waters warmed to bathtub temperatures, the schools moved into the cooler water below the thermocline. The geographical distribution of sharks is also affected by temperature. Reef sharks (members of the family Carcharhindae), and hammerhead sharks (family Sphyrnidae) migrate into the temperate latitudes during the summer when the local waters are warmer, but they return to the warmer latitudes during winter. For example, scalloped hammerhead sharks enter the Gulf of California in late spring and stay until late fall—during winter the water temperatures are below 70°, the lower thermal limit of many of these species. During warm summers, solitary scalloped hammerhead sharks occasionally venture as far north as San Diego during the late summer when the local waters are warmest.

Throughout the day, Steve and Scott would take turns swimming down to photograph the individuals within the schools, using the stereocamera. They would carry the apparatus down until level with the school of hammerhead sharks. The aluminum beam was held so that it was parallel to the long body of the nearest shark and the handle and trigger squeezed to take two photos.

At the same time, I swam down repeatedly to the edge of the school with this massive underwater housing. There, I remained suspended in one place, barely kicking, or kicking more rapidly to propel myself forward with the same speed as members of the school were swimming. The housing was cumbersome, weighing over 50 pounds, as might be expected of the first underwater housing made that contained both a camera and a recording deck. The housing had a cylindrical Plexiglass port bolted on its front, which bristled with knobs to adjust the brightness of the picture or to zoom in on a particular shark performing an interesting behavior.

What exercise this diving was! Over and over, I pushed this piano-like underwater housing to depths of 60 to 70 feet, stayed there a minute and a half filming the group, and expelled air as I slowly rose to the surface. This was grueling physical work, yet it was very exciting being among the sharks hour after hour, day after day.

Returning to camp, we would then walk across the beach to the three brothers, who stood next to separate piles of sharks. At this time, they would be quickly and efficiently cutting fillets from the sharks that they had caught that day. I would stretch a metric tape measure from the tip of the snout of each shark to the end of its tail. This measurement of size is called a Total Length. I would then spread apart the two arms of a pair of calipers so that one was even with the tip of a clasper and the other with the junction of its base with the shark's abdomen. The tips of the calipers would then be placed on the tape measure to record the length of the clasper. Male maturity was determined in the following way: A graph was drawn with clasper length on the vertical or x-axis and shark length on the horizontal or y-axis. I then examined the points on this graph, indicating the sizes of the claspers for sharks of differing lengths. These points form a slanted line rising upward: the claspers become larger as the shark grows longer. However, at an intermediate length, the points indicating the size of clasper do not always fit on the slanted line; some are above it. This large increase in clasper size corresponding to a small increase in shark length indicates that the shark's clasper grew more rapidly than the shark. This is similar to puberty in human females, when the breasts develop faster than the female becomes taller. This spurt of growth in the male clasper is accompanied by a hardening of the cartilage to make the clasper rigid. The least equivocal indicator of reproductive readiness is the presence of spermatophores, or little packets full of sperm, in the canal passing through the middle of the clasper. I looked for all of these indicators of maturity in males. I found that male hammerheads first became reproductively mature at a length of 170 centimeters, or 5 feet 7 inches.

The method used to determine reproductive readiness of females was easier. I removed the uterus from each female on the beach and examined it closely for ova, or eggs. There would be ova of all sizes in a mature female. I then moved the two legs of the calipers together so they touched the top and bottoms of each egg. I later would perform the same analysis as I had done with the males, plotting ovum size on the x-axis and female length on the y-axis. At the smaller lengths of the female, the points indicating ovum diameter would be at the bottom of the graph corresponding to small diameters. At a threshold length, there should be a scatter of points, some of small sizes but many others of a large size that indicated that these females had become reproductively mature. I found that female hammerheads first became reproductively mature at a length of 214 centimeters, or 7 feet. This was curious: females became reproductively mature a foot and a half larger than males. I wondered why this was the case.

We waited until rough seas kept us at home to analyze the stereophotos. At this time, Steve and Scott would spend their time viewing the photographs under a microscope. A scale was superimposed on the photographs using a mirrorlike device called a camera lucida. The same hammerhead was visible on both photographs, but its position was different on the two photographs. This was obvious if you placed two slides on top of each other. The snout of the hammerhead on the photograph from the camera without the viewfinder was a little to one side of the snout of the hammerhead centered in the photograph using the viewfinder. One could measure this small distance by superimposing a scale with the camera lucida, and this small reference distance was equal to the 50-centimeter distance between the two cameras. They next used the microscope to measure the minute distance between the snout and tail on the shark's photographic image. The length of the shark was then determined by dividing the distance between snout and tail by the separation between the shark's snout on the two photographic images and multiplying it by the 50-centimeter camera separation distance.

For example, this reference distance fit exactly seven times into the image of the largest shark ever measured within a school. This was a massive female at the center of a school swimming over El Bajo Gorda, a church spire–like pinnacle located a few miles off the tip of the Baja Pensinsula. The length of the shark was then 350 centimeters or seven times the 50-centimeter separation distance between the cameras. In reality, a small correction was made for a slight convergence of the optical axes resulting in her length being even longer, 371 centimeters, or 12 feet 2 inches, twice my height. This was one big female shark!

However, such large females were uncommon in the schools. More than three-quarters of females measured with the stereocamera within the schools were smaller than the 219-centimeter or 7-foot length at which females became reproductively mature. The preponderance of immature females seemed contrary to the idea that the groups of females were mating aggregations.

Females were much more common than males in schools. During three successive years, I made extended dives in order to determine the sex of the hammerheads within the school. My practice was to swim down through the school while kicking vigorously, then hold my legs still and slowly float upward while looking on the underside of each shark passed to see whether it had claspers or not. At times, I increased the depths of my dives by using a small "pony" bottle, carried on my back as an air supply. I could continue to take breaths while descending without filling up my lungs with air, because the air already present in my lungs compressed as I swam deeper and encountered increasing pressures because of the added weight of the water. Reaching the bottom, a broad plateau, 120 feet deep, from which the seamount pinnacles rose to near the surface, I began to rise slowly to the surface, exhaling as I passed through the school of hammerhead sharks.

Don Nelson and I counted 155 females to only 57 males during the first year (1979) at the seamount. During the second and third years (1980 and 1981), spot checks by myself yielded 72 females to 13 males and 120 females to 20 males. The longer I thought about it, the less convinced I was that that the preponderance of one sex within the schools and the immature reproductive state of the majority of females ruled out reproduction as a function of the schools. Perhaps females were competing among each other for the apparently rare males. Could these dominant females be competing for a favorable position within the school, where they would receive more attention from males?

While Scott and Steve tediously recorded the sizes of sharks within schools at our camp at Isla Pardito, I viewed hours and hours of video of the behavior of sharks within the schools. Perhaps the videos would yield more insight into the function of the schools. From the videos, I quickly found out that the schools were not peaceful at all. To start with, many of the females had small oval abrasions on their heads, where their sandpaper-like skin had been rubbed off. The scars were initially white and became black as they began to heal. The same scars were visible on the underside of the snout near the mouth of other females. These scars were probably not the result of male love bites. Three-quarters of the sharks measured with these scars were less than 7 feet long, the size at which females became sexually active. Males shouldn't be motivated to bite these prepubescent females.

I noticed from the videos that occasionally a shark would accelerate downward and strike her nearest neighbor with her lower jaw just to the side and in front of her first dorsal fin. This action appeared to be the cause of the abrasions. These aggressive sharks seemed to be females that were larger than other school members, but it was hard to know this for certain because they remained only momentarily in view as the school moved out of sight. Sharks came into contact with others, causing these abrasions, most often after performing the acrobatic *Corkscrew*. A shark would dash into a tight looping trajectory while twisting its torso completely and then contact the shark below with the underside of its snout, leaving a white contusion. The video was a useful tool because it permitted me to view the behavior in slow motion over and over again. The behavior resembled the reverse somersault and full twist performed by platform divers at swimming meets. Only by using the video could one see that the shark often came in contact with another. One could also look on the video for the shark's claspers and determine its sex. From the videos, I was able to determine that most of the sharks performing these two behaviors, *Corkscrew* and *Hit*, were females.

Whenever a female performed one of these behaviors, the females around them darted off in various directions. I called this behavior *Acceleration*. It consisted of a rapid burst of swimming with rapid shaking of the head. Other females performed another behavior, *Head Shake*, consisting of the hammerhead swinging its head back and forth over a wide arc several times in quick succession while suddenly propelling itself forward with a rapid acceleration.

I made a flow diagram of the sequence of the behaviors. Engineers often resort to this step-by-step analysis to document the successive steps in computer programs. My diagram indicated how often one behavior followed an earlier one by the thickness of the arrow, drawn between symbols for the two behaviors. Perhaps this technique would provide a rationale for why the sharks performed these behaviors in the schools.

The rationale became apparent quickly. The *Hit* was often followed by the recipient and other adjacent sharks scattering while performing *Acceleration* and sometimes *Head Shake*. The *Corkscrew* was followed by the same two behaviors, but less often. I reasoned that the *Corkscrew* was a signal because, like *Hit*, it caused other sharks to leave the comfort zone, the center of the school, but unlike *Hit*, the *Corkscrew* avoided body contact and the possibility of injury. The sharks performing *Acceleration* were usually smaller than the sharks displaying the *Corkscrew*. There was less of a difference in the sizes of the sharks performing the *Corkscrew* and the *Head Shake*. According to ethological thinking, the *Head Shake* was likely to have been motivated by the conflict between attack and escape (fight or flight). This might occur if the shark was not able to easily determine whether its aggressive neighbor was larger. The head of the shark would alternatively be swung toward (reflecting a motivation for attack) and away (showing an inclination to flee) from the slightly larger aggressor as it swam toward the edge of the school.

Torso Thrust

A male hammerhead shown exhibiting the *Torso Thrust*, which consists of thrusting his midsection to the side while rotating his clasper several times to catch the attention of a dominant female swimming in the center of the school. When the maneuver is successful, the female joins the male and they leave the school to mate at a later time. (COURTESY OF BLACKWELL PUBLISHING)

It appeared that the females within the schools were fighting among themselves for a favored central position. But why do this? I noticed that males often swam into the center of the school with explosive vigor. As the male entered the school, he beat his tail to one side, propelling his forward body to the other side in a jerky movement. The male tilted slightly and rotated its clasper to the side at the same time. Could these males be seeking to pair with the large dominant females situated in the center of the schools? Behaviors that served as signals often became exaggerated versions of behaviors used in other activities. Could rotation of the clasper, which was used to fill the siphon sac with water, have become a male advertisement directed toward females to interest them in mating? The male sharks that performed *Torso Thrust* appeared to pick out a female, swim alongside of her, and leave the school together. However, to my frustration, I never saw them mate at the seamount, but mating has since been observed there. The male hammerhead accelerates to the side of the female that he has paired with, seizes her pectoral fin, rolls over and bends his midsection toward her to insert his clasper into her vent. Once the male inserts his clasper, the coupled sharks stop swimming and remain still while they slowly sink to the bottom.

If school members were fighting for a central position within the school, and success in these contests depended on the relative size of the combatants, one would expect to find larger sharks closer to the school's center, with the largest ones in the center. This was a prediction whose validation would give credence to the existence of a dominance hierarchy within the group. How was I going to determine this? The stereocamera could be used to answer this question too. By making a few additional measurements on the photos, I could determine where any point on an image on a photograph was in a three-axis coordinate system with the camera in its center. This consisted of x- and y-axes at right angles to each other on a flat horizontal plane centered in the camera lens and a vertical z-axis, intercepting the other two axes in the lens and leading up and down in the water column.

It was essential to know whether sharks at the top of the school were smaller than individuals in the school's interior or whether sharks on the edges were smaller than others near the school's center. First, I descended to a position near the top of the school and took pairs of stereophotos pointing the camera lens downward toward the center of the school. Second, I swam to a position alongside of the school and took pairs of stereophotos with the camera pointed directly into the school's center. We then subtracted the distance from the camera to the outermost shark from the distance to a particular shark to determine that shark's position relative to the school.

We found that it didn't matter whether the distance into the school was from the top or side, hammerhead lengths increased by 16 centimeters, or 6 inches, for every meter into a school. This was a substantial increase in size toward the school's center: a shark at the center of a school ten sharks wide could be 2 1/2 feet longer than a shark at the edge of the school. Everything seemed to make sense now. By chasing away smaller sharks, the large female sharks were maintaining a position in the center of the schools, where they were singled out by male sharks. This was a complicated mating ritual. This matching of the largest females with males of the species appears to be the function of the schools.

The massing of females into huge schools at the seamount struck me as a curious phenomenon. Why didn't more males join the groups?

And where were those large males? To start with, hammerhead mothers had the same number of male and female pups. The fishermen told us about the depths at which they set their nets to catch hammerhead sharks. Many males, but few females, were captured in less than 50 meters of water. This indicated that the males preferred to stay in shallow water close to shore, perhaps where there was less risk of being eaten by larger sharks. But females of the same size must have moved offshore where food would be more abundant.

I examined the stomachs of the hammerhead sharks caught by the three brothers to determine what they were eating. This was dirty work. I cut open the stomach with a knife and placed all of the contents in plastic bags. Later, I determined the number of stomachs with each type of prey, counted the items, and measured the volume of each item. This I determined by placing the item in a beaker with a volumetric scale on the side, which was partly filled with water, and noting the change in the level of water. All of these indicators were combined to give an index of dietary priorities.

I usually knew the habitat of the prey items from other researchers' observations. For example, many squids spent the night swimming near the surface, yet lived during the day at great depths. The female hammerheads caught by the fishermen and similar in size to the smaller individuals present at the seamount had stomachs full of the parrotlike beaks of deepwater squids. The stomachs of males of the same size were filled with bottom fish, indicating that males this size stayed near shore. Most significant was that the stomachs of the smaller females, caught offshore, had fuller stomachs than males, indicating that the females had greater success in feeding after their offshore migration.

It dawned on me that the females might be moving offshore and feeding more in order to grow faster than the males. As mentioned, female hammerheads become mature at 1 1/2 feet larger than males, 7 feet versus 5 feet 7 inches. A colleague of mine, Frank Schwartz, had found that females grow faster than males. He measured lengths of hammerhead sharks and related these to age, estimated from the number of concentric rings on their vertebral elements. The seasonal change in water temperature governs the individual's rate of feeding throughout the year, resulting in a yearly gradient in the deposition of material within each ring. The rings on an element of a shark's backbone are similar in appearance to tree rings.

Females probably move into a different habitat to grow faster than males because they need to be larger than males to successfully reproduce. More energy must be available to nourish embryos that can occupy a third of their body mass, in contrast to the miniscule investment of a few sperm from the male. Perhaps this is also the reason why females continue to grow larger after they become mature, while males grow only a little more. It is reproductively advantageous to be a larger female. The number of pups that a female has is directly related to her size. The smallest females may have only a dozen pups, while the largest may have up to forty pups per litter.

I suspect that female scalloped hammerhead sharks segregate from the males in order to grow faster. They thus can reach the larger size needed to reproduce at roughly the same age as the males and match their reproductive output throughout their life to that of the males. Consistent with this theory is that the few sharks that lay eggs, which require less energy to develop than shark embryos, segregate by sex less commonly. The scalloped hammerhead is a curious species. The anatomy, physiology, and behavior of female hammerheads are so dramatically different from those of the males that they barely seem to belong to the same species. And the female, while perhaps not "more deadly than the male" as the old saying has it, is certainly the key to the riddle of schooling behavior in the species.

CHAPTER 7

SHARK RUSH HOUR AT GORDA SEAMOUNT

I made four seasonal expeditions to the Gulf of California between November 1980 and July 1981. During each trip, I visited six locations where hammerhead schools had been observed along 250 miles of coastline extending southward from the town of Loreto, two-thirds of the way down the thumb-shaped Baja Peninsula to San José del Cabo, at the end of the thumb, near Cabo San Lucas. Our itinerary included visits to two seamounts, Espíritu Santo, which was closest to La Paz, and Gorda, which was nearest to San José del Cabo. Of the four islands where we worked, two, Danzante and El Solitario, were close to Loreto, and two, Las Animas and Cerralvo, were near La Paz. I made day-trips to the hammerhead sites near Loreto aboard my panga. We visited the study sites near La Paz on ten-day cruises aboard the *Juan de Dios Bátiz* of the CICIMAR laboratory in La Paz.

It was a sunny but cool day on May 6, 1981, when we visited the majestic Gorda Seamount aboard the *Juan de Dios Bátiz* during our spring expedition to look for scalloped hammerhead sharks. It was the next to the last of my years as a graduate student at Scripps. Gorda Seamount, situated six miles off the southern tip of the Baja Peninsula, was the southernmost place where we might see sharks. The seamount rose up from the bottom to form a pinnacle like a church spire at 120 feet deep—too deep to be seen from the surface. The sea surface that day was flat, but moved up and down slowly as large rounded waves, half the length of the boat, passed by on their way from the south. The round-bottomed *Juan de Dios Bátiz* rolled back and forth in the large swell, which came from a storm somewhere far south of us. Storms (or the dreaded hurricanes, called chubascos) here were generally during late summer and early fall and were uncommon during winter and early spring. Perhaps this swell from elsewhere announced the arrival of summer and the hammerhead sharks to the Gulf of California.

During the prior week, we had visited three other favorite abodes of hammerheads in the lower third, or southernmost region of the gulf, where it opened into the eastern Pacific. The northernmost place visited was Las Animas, a huge and imposing rock jutting out of the sea a dozen miles off the coast of San José Island, which in turn was 10 miles from the mainland. We had slowly swum around the coast of this island, intently searching for the hammerheads without seeing a single one. In the summertime, the water was alive with dolphins, birds, fishes, and, of course, hammerhead sharks. Our next stop on our way south was the Espíritu Santo seamount, where we saw fewer than half a dozen hammerheads, all swimming either singly or in pairs. Little else was present there: no diffuse schools of green jacks and pompano and no tunas and snappers.

The third place we visited during our expedition was Las Arenitas, a rocky reef near the cluster of rocks, separated by not more than 200 yards from a few small sandy beaches along the coast of Cerralvo Island. The ocean side of the reef sloped steeply into deep water. Here we had first seen schools of hammerhead sharks two years earlier, but now we found only small, loose aggregations of oval-shaped creolefish at the edge of the reef.

The *Juan de Dios Bátiz* took almost a day, driving southward, to reach the long, wide beach of San José del Cabo, cluttered with the pangas of the town's shark fishermen. Here the large surf prevented fishermen from keeping their pangas at the edge of the water. I marveled at how the local fishermen beached their boats full of sharks caught at the Gorda seamount. As they approached shore, the skippers gripped onto the arm of their outboards and twisted the throttle.

This made them go faster and lifted the bow of the boats. Just at the moment that they arrived at shore, the fishermen would turn off the motor and lift its outdrive so that the boat shot up onto the beach. The other fishermen, upon seeing this, slowly shuffled over to boat. Two or three fishermen would obediently grasp the gunnels, the flat lip around the side of the boat, lift the fisherman's boat up and carry it up the beach to safety. These precadores regularly set their long lines and nets at both the inner and outer Gorda seamounts. The inner mount was 6 miles from shore, the outer twice that distance from the coast. The skipper anchored the boat close to the beach, where we spent the night before driving to the inner Gorda seamount early on the following morning.

We took a little over an hour to reach the general area of the seamount, which we found by moving offshore until the top of a hill on the coast lined up with the edge of a mountain range farther ashore. The captain kept these landmarks aligned, one in front of the other, while proceeding offshore until a second set of landmarks became aligned. This was the best way of locating a place before the Global Positioning System was available. We then got into the small panga, turned on the fathometer, made sure that one pair of the land-marks was aligned, and again motored the boat until the other land-marks became aligned. There on the chart was a black trace indicating the shape of the seamount: a black line on the sonar's paper suddenly jutted upward to a point and quickly dropped downward. I lifted a small anchor out of the boat, let out line until it reached the bottom (evident by the loosening of the line), quickly attached a buoy, and threw it in the water. Now we had an object on the surface over the seamount that would help us locate the steep pinnacle again and pos-sibly the hammerheads.

The helmsman then drove the panga up to the buoy and carefully looked for the swirl or eddy of water caused by the water flowing past the stationary buoy. The direction and amount of turbulence provided a rough estimate of the direction and speed of the current at this time of the day. He watched some bubbles travel with the current. It was moderately swift, perhaps a meter and a half per second, close to the limit of our ability to swim against the current. We would have to be driven up-current of the seamount and dropped in the water so that we could drift over the pinnacle, looking for the sharks.

Strong currents are often present around seamounts, which is not surprising if you think about it. A mass of water, moving slowly over the bottom far below, speeds up as the same amount of water passes through a smaller space between the surface of the sea and the shallow seamount. The end result, of course, is that roughly the same amount of water passes over the bottom in both places, but at different speeds. This same phenomenon explains why winds are so strong at mountain passes. Air must pass through a smaller space between the mountain and the earth's upper atmosphere.

We let ourselves into the water, wearing our masks and snorkels, and started swimming over the seamount in the usual formation. Each of the three of us kept within view of our nearest buddy and roughly abreast of each other in a line perpendicular to the direction of the buoy. Searching for hammerhead sharks at the Gorda Seamount (El Bajo Gorda in Spanish) was difficult and hazardous. To start with, you could never see this seamount from the surface. Diving without the bottom in sight makes anyone feel uncomfortable and disoriented. This was an eerie experience, to which we had to become accustomed with time and experience.

We used the staccato *clickity, click, click, click* of snapping shrimp, hiding below in the rocks of the seamount, to locate it. This cacophony became louder as we drifted directly over the seamount. Less sound energy was absorbed when the sound traveled a shorter distance to the surface. However, despite having this cue, we could miss the pinnacle because the peak was so pointed, abruptly rising from surrounding depths of 200 to 300 hundred feet. The currents flowed very fast, and we could drift pass the seamount before making a dive. That meant we would fail to see the hammerheads because we were down-current of where they were—directly above the peak of the seamount.

There were often huge rolling waves at El Bajo Gorda like the ones present that particular day. My greatest fear when working here and at El Bajo Espíritu Santo was that someone might be swept out to sea in the current and be lost, since there was no land south of us for thousands of miles. At the start of each expedition, I solemnly explained to everybody the importance of staying together when drifting over the seamount and searching for the hammerheads. It was the duty of the helmsman on the panga to keep track of us at the surface of the water. That person tried to maintain eye contact with the divers as much of the time as possible. Each of us, taking turns as helmsman, became very frustrated today. We could see the team of divers only briefly when on the crest of a wave and not at all when in the valley of the same wave. Toward the end of the study, we carried whistles on our wrists to use to make a loud sound, when momentarily lost, in order to alert the helmsman of our position. We drifted by the seamount six times this day without seeing any fishes. There was no sign of the large school of hammerheads that occupied the seamount last summer.

Our next objective was to make a scuba dive over the seamount and to look for the sharks. We could tell from the eddy of water around the buoy that the surface current had slowed down. Two of us, with regulators and tanks, jumped in the water next to the buoy, put our mouthpieces in our mouths, and slowly swam down the line until we were 20 feet from the bottom. "Brrr," I thought, "the water is cold down here over the seamount." We were now 120 feet deep and almost level with the highest point of the seamount, not far away and in front of us. We began vigorously kicking our legs, propelling ourselves forward against the swift current. Our progress over the bottom, visible below, was agonizingly slow because the current was now so strong. Our plan was to first swim over the pinnacle, not far from us, and then drift back over the seamount. We hoped to catch a glimpse of a school of hammerheads either up- or down-current of the pinnacle.

Looking over at my partner, Bob Butler, I noticed he was kicking much faster than I was. "Aha", I thought, "I had better lend him my extra pair of long fins for tomorrow's dive." He could then better deal with the relentless opposition to forward movement produced by the strong local currents. However, his exertions did not bother me at the time because he was not only in peak condition physically, but also highly experienced, the two ingredients that make a first-rate diver. In fact, he had made over a thousand scuba dives in the waters off San Diego, assisting Mia Tegner, a colleague of mine at Scripps in studying the ecology of kelp forests.

Once we reached the pinnacle of the seamount, he unexpectedly left my side and started swimming rapidly toward the surface. This struck me as odd because we both surely had enough air to stay underwater longer. I waved at him with one hand and pointed at the seamount with my other indicating that he should come back. However, he kept swimming upward despite seeing me. His rate of ascent seemed much too fast to me. I was concerned that such a quick ascent would cause the air in his lungs to expand too rapidly, causing tissue damage, a diving malady called chest embolism. I took off after him, and after a few seconds was able to grab him by the ankle. He then stared at me with a perplexed expression on his face. We swam a little farther forward, with me looking to either side for the hammerheads, before I decided that something might be wrong with him. He wasn't looking around for the hammerheads like me. I thought that we had better abort the dive and swim to the surface for the sake of his safety. I wasn't sure he was in trouble, but reasoned that it was better to be overly cautious when diving in these perilous currents. We stayed five minutes just below the surface to permit the nitrogen, absorbed in our tissues at greater depths, to dissolve from our bodies. He seemed much better now. We then rose to the surface, where the helmsman on the panga was waiting for us.

As soon as he was in the boat, he exclaimed that the constant exertion required to swim against the strong current had tired him and he had become claustrophobic. We agreed that he had the symptom called tunnel vision, caused by the build up of dissolved carbon dioxide in the blood. If the concentration of carbon dioxide becomes too high and the oxygen level too low from overexertion, one can even pass out underwater.

This incident, although minor, emphasized again to me the need to take all possible precautions to avoid accidents while diving during the cruise. Furthermore, I suspected that we had been competing to appear the stronger diver, and this may have partly caused the problem. I resolved to do my best to avoid any tendency toward competition between myself and other cruise members in the future. We returned to the seamount on the following day and saw only a single female hammerhead swimming among a single small school of oval silver fish with yellow tails, the yellowtail jack. This place, which had been alive with fish the summer before, was now as devoid of animal life as a desert during the hottest time of the day.

I remember vividly our next trip to Gorda Seamount, although over twenty years ago. It was on the thirteenth of May, only seven days later. I had not intended to return so soon, but Ciro, the cook aboard the *Juan de Dios Bátiz*, informed me the evening before that the hammerheads had returned to the inner and outer Gorda seamounts. His brother, who lived in the town of San José del Cabo, had called earlier during the day to say that he and other fishermen had caught many hammerheads in their nets earlier that day at Gorda. He went on to say that Pedro, the American shark man (me), Felipe Galvan-Magaña, and the other students at CICIMAR, the laboratory for which Ciro worked, should come quickly. We drove to San José that evening and hired a local fisherman and his son to take us to the seamount on the following morning. We arrived at Gorda to see an armada of the long, slender fiberglass outboard pangas scattered over the horizon with two or three persons in each fishing for snappers (*huachinanga* in Spanish) and sharks. What a radical change from a week ago when the *Juan de Dios Bátiz* and my panga were the only boats present at Gorda!

The fisherman found the seamount using the same ranges that we had used to find it. His son lowered a rusty pipe wrench with thin monofilament line leading to a plastic gallon milk carton that floated at the surface. This was the local fishermen's less high-tech but equally effective version of our fancy mooring, made of a Dansforth anchor, nylon line, and a large tear-shaped inflatable buoy. The fishermen drove the panga a couple of hundred feet up-current of the buoy. Here Felipe and I, sitting on either side of the boat, simultaneously fell backward into the water wearing our masks, snorkels, and fins. When the bubbles from my entry into the water dispersed, I was dumbfounded by the concentration of fish around us. Just a week before, there had been nothing here.

It seemed like we had joined a piscine version of the rush hour at a train station. The difference between now and then was like the contrast between an empty stadium and one crowded with thousands of supporters cheering on their team. There were fish everywhere! Around me was a school of small hammerheads between 4 and 5 feet long. I counted 125 sharks in view while slowly rotating in a circle. The sharks at the top of the school were close enough to touch from the surface. Just beyond the school of sharks and to the side was a massive school of silver and blue-striped skipjack tuna. The individuals, each the size of a small football, were more tightly packed in this school, and the members seemed to act more orderly than the sharks. Two blacktip reef sharks darted back and forth between us and the school. The school was so long that it extended as far as I could see— visibility was over 100 feet in the crystal-clear water at Gorda on that day. Below us was a cyclone-shaped school of large gray-striped snappers, each 2 feet long, slowly swirling around in a circle. Everywhere were small green jacks and pompano, darting here and there to feed on small shrimps, larvaceans, and other members of the plankton.

As we floated by the buoy, an even larger and more compact school of red-hued snappers came into view below. There underneath them we could see the rocky surface of the seamount. What remarkable water clarity! We must have been able to see farther than 120 feet that day. Near the surface, there was a massive round school of silver plate-shaped crevalle jack. To the side was a tightly packed school of yellow-colored amberjack, each member probably weighing 20 to 30 pounds, that also extended outside our view. Cruising at the surface was a long slender silky shark. Swimming slowly beneath was a huge shark with a long tail. Its tail was that of a thresher shark, but it was impossible to see whether it was the shallow-living common or deep-dwelling big-eye thresher. The place was just alive with fish. We spent the morning taking stereophotographs of members of the hammerhead school to later determine their lengths. I also tagged two individuals with spaghetti-type tags to aid in identifying them in the future.

We came back to Gorda on the following day. The school of hammerheads was now so large that it was impossible to count the members. Instead, I looked at the second hand on my dive watch and timed the passage of the school. It took more than 120 seconds, or 2 minutes, to pass me. There were now probably two or three hundred hammerheads in the school. The two schools of snappers had now grown to twice their size during the previous day. Joining the massive schools of skipjack and amberjack was a third school, composed of yellowfin tuna, each member probably weighing 40 to 80 pounds.

During a scuba dive, we observed several tongues of dark material projecting upward from the depths around the seamount. We swam over to one of these strata and found it to be composed of thousands of tiny larval fishes, each no longer than your smallest finger. Two large manta rays appeared from below. Their paddle-shaped appendages, rolled into a scroll extending outward on either side of the head, unraveled and came together below the mouth to produce a biological funnel that forced more plankton into their mouths. The mantas swam slowly back and forth through these tongues of larval fishes, gorging themselves with food. Then above us appeared a whale shark, a dull-green-colored female with bright blue iridescent spots. She swam just below the panga, visible above us at the surface, and appeared to be 25 to 30 feet long. Her mouth was open wide because she had come to share the feast of larval fishes with the mantas.

My observations on the second day made me wonder whether more hammerheads, tunas, and snappers were arriving at Gorda each day. Were we witnessing a massive immigration of species into the Gulf of California? Was each species coming from a distant and separate locality? Or were all of these species migrating in a coordinated and related group of species, which originated from the same locality? This seemed to me a more parsimonious (or simpler) explanation for the arrival of so many species at a single seamount at the same time. If they were to assemble from widely separated geographic locations, they would have to regulate their departure time and rate of movement over the seafloor in order to arrive simultaneously with other species, no simple task in an apparently featureless ocean. The various species would also have to compensate for the different local currents that might cause a delay by deflecting them from their intended course. Was not a simpler explanation that they all moved together as a group? As easy as it was to believe that large fishes such as the hammerhead shark and manta ray migrated over large distances, it was

harder to believe that small species such as the green jack and pompano might keep up with them during such a migration. Yet the presence of small species may be the reason the larger predatory species
join the assemblage, because the former are forage.

And why were all of these species assembling at the seamount
now? A storm had come to Gorda between our visits in the *Juan de
Dios Bátiz* and the local fisherman's panga. Had this storm brought
with it the first warm water of the year into the Gulf of California?
Certainly, we had experienced warmer water and higher water clarity
after the storm. We needed now to see some satellite images of the
local sea-surface temperatures and chlorophyll concentrations before,
during, and after the storm. The concentration of chlorophyll pigments was an indicator of the local density of tiny microscopic plants
(phytoplankton) in the surface waters. I couldn't wait until we
returned to Scripps to get some images from the new Satellite Facility.
My impression was that we had witnessed an immensely important
event, a massive immigration of species into the gulf, and it was now
crucial to learn why it occurred at this time of the year and at this
seamount.

Back at Scripps, Steven Butler, the graduate student working with
me at the time, began searching the archive of satellite images at the
Satellite Facility for images taken during the two days we visited
aboard the *Juan de Dios Bátiz* and the fisherman's panga. He substi-
tuted a nocturnal work schedule for his customary daytime one so
that we could purchase image-processing time at a bargain rate. He
spent many a late night alone sitting in front of a computer and viewing colored satellite images on its screen.

The National Oceanic and Atmospheric Administration, or
NOAA for short, had launched a satellite that circled the earth twice
daily and was equipped with an electronic sensor that measured ocean
temperature by recording the amount of heat radiated from the thin
veneer of water on the surface. The sensor was called the Advanced
Very High Resolution Radiometer or AVHRR for short. Its spatial resolution was not actually that high, averaging energy in the infrared
band of the electromagnetic spectrum, radiated from a square of sea
surface with its sides slightly over a half mile long.

NOAA had launched a second satellite with an electronic sensor, the Coastal Zone Color Scanner (CZCS), that detected wavelengths in the green region of the spectrum that radiated from the sea surface. The concentration of phytoplankton, microscopic algae floating at the surface, is proportional to the amount of light emitted from their green chlorophyll pigments. All plants, marine or terrestrial, use these pigments to convert the energy in sunlight to organic matter using carbon dioxide, certain inorganic nutrients such as nitrogen and phosphorous, and water as additional ingredients. Phytoplankton, which are microscopic plants, comprise the lowest level of the trophic, or food, pyramid in the oceans. The abundance of every species, whether a creolefish feeding on animal plankton or hammerhead shark eating a creolefish, is related to the amount of phytoplankton. This, in turn, is related to the concentration of chlorophyll in the water. This relationship exists because large predatory species feed on smaller predators, and these eat filter-feeding animals (larval fish, larvaceans, jellies), and these in turn consume the microscopic algae. Hence, the concentration of chlorophyll is considered by oceanographers to be an indicator of the richness in life, or productivity, of a particular spot in the ocean.

Steve and I were impressed by the difference between the images of sea-surface temperature and chlorophyll concentration before and after the assemblage of fish arrived. We had to use satellite images from the day prior to our initial visit to Gorda because clouds obscured the sea surface during both of the days we were there. Satellite images use a color spectrum to indicate the range of measurements. We denoted cold water by the color blue, cool water by green, warm by yellow, and hot by red. On this thermal image, Gorda was within a green crescent of cool water. The crescent started to the east of the "fingernail" of the thumb-shaped peninsula, passed west along the tip and toward the nail, extended southward at the middle of the tip, and curved eastward to complete the crescent. We chose dark blue to indicate a low concentration of chlorophyll, light blue and green to denote intermediate amounts, and yellow and red to designate high concentrations. On the chlorophyll image, Gorda was within a large light blue triangle, indicating a substantial amount of phytoplankton was in the area. This wedge of cool, plankton-rich water is usually present off the tip of the peninsula from October to April and is due to the steady winds from the north causing water to rise to the surface that is rich in nutrients, which permit phytoplankton to grow.

We now looked at a pair of images taken by the satellites when we visited Gorda again a week later. On the AVHRR temperature image, the green band of cold water was no longer present at the tip of the peninsula. A green band now extended northward around the tip of the peninsula as though cooler water had been pushed into the gulf by the entry of warm water. The tip of the peninsula was now surrounded by yellow, indicating that Gorda was now bathed in warm water. Furthermore, on the CZCS chlorophyll image, the area over the seamount was a darker blue, indicating lower concentrations of plankton. The change in local conditions apparent on the large-scale satellite image was consistent with our observations of warmer water and better clarity present at Gorda when we returned and found the massive assemblage of fishes.

Could the hammerhead sharks and other assemblage members have migrated to Gorda within a mass of warm water that originated south of the Gulf of California? Favoring this hypothesis was that few assemblage members were observed during our fall, winter, and early spring expeditions to favorite hammerhead abodes inside the gulf along the roughly 250-mile latitudinal gradient northward along the Baja Peninsula. Hammerheads and silky sharks, two assemblage members, are caught during the winter farther south, outside the Gulf of California and along the southwestern coast of Mexico by fishermen based at Mazatlán. The water in this region is considerably warmer throughout the winter than the waters of the gulf.

During our fall expedition aboard the *Juan de Dios Bátiz* in late October 1980, the waters in the lower gulf had still been warm, but were beginning to cool as cold fronts periodically descended the length of the Baja Peninsula, bringing with them strong winds and rough seas. At that time, we saw fewer than a dozen hammerhead sharks at the four hammerhead locales—Las Animas, Espíritu Santo, Las Arenitas, and Gorda. There were no aggregations of green jacks and pompanos, pairs of manta rays, or whale sharks feeding on plankton at the locales. The dense schools of tunas and snappers were nowhere to be seen. On the way back from Las Arenitas, rough seas forced us to stay within a sheltered bay on the island of Espíritu Santo for three days until the front passed and the seas subsided. Each day, we would pull the anchor, start to motor out to the Espíritu Santo seamount, only to turn back because of the rough seas. The strong winds propagated large waves in a direction opposite to that of the tide, resulting in the mariner's worst nightmare: turbulent seas comprised of waves with high crests and short wave lengths. Furthermore, when these waves broke on the steep cliffs of the many local islands, they were reflected away from the islands in various directions. These reflected waves added to or subtracted from the heights of the other waves and made the gulf one jumbled mess of waves.

That particular fall expedition was notable not for the few fishes we observed in the gulf, but for what happened to Eduardo, the captain of the *Juan de Dios Bátiz*. We were leaving a third time for the Espíritu Santo seamount when he rushed up to me, eyes wide with pain and fear, speaking hurriedly in Spanish. He pointed repeatedly at his throat with one hand and then toward the plastic gallon milk jug held in his other hand. I quickly surmised that he had drunk something very toxic that was contained in the bottle. I grabbed the bottle from his hand, held it close to my nose, and sniffed. "Ouch!" I exclaimed. The odor was overwhelming. The bottle contained formalin, a concentrated solution of water and formaldahyde. This chemical is extremely poisonous and, mixed with water, kills bacteria and preserves whatever animals are kept in the mixture. It was a hot day and the captain was very thirsty. He grabbed the unlabeled plastic container, in which the Mexican students kept their formalin. Thinking that it held water, he had held it up to his mouth and swallowed a goodly amount of its noxious contents before he realized his terrible mistake.

The corrosive fluid was now eating away at his throat, stomach, and intestines. I ran around the boat looking for the cook, and upon finding him, asked for a carton of eggs. I broke them one by one, emptying the raw egg whites and yokes into a cup, and gave this *huevo* cocktail to the captain to drink. It took nine raw eggs to make him finally vomit. We then helped him into the panga, in which we drove him thirty-five miles to La Paz. During most of the trip, the captain continued to throw up formalin mixed with raw eggs. He must have drunk a considerable amount of formaldahyde because the water in the bottom of the boat, mixed with the vomit, stung our feet and smelled strongly of formalin on the way to La Paz. When we arrived onshore, my colleague Henk Nienhuis, an unflappable Dutchman, hailed a taxicab, told the driver that the captain was terribly sick, and asked him to take us to the hospital. The cabby then drove off like a madman, narrowly missing cars coming toward us in the opposite direction as he accelerated around the many turns leading from Pichilingue, where our boat ramp was located, to downtown La Paz. Halfway there, Henk looked at me, winked, and whispered into my ear that we might all end up in hospital beds. After talking with the doctor at the hospital, the captain suddenly appeared to feel much better. The doctor informed us that, if blood was absent from his vomit (which was the case), the formalin had not punctured his stomach and intestines and he would recover relatively quickly. But he would have to abstain from eating chiles for the next six months. We dropped him off near his home later that day, much relieved that the captain had survived his terrible ordeal.

During our winter expedition in January 1981, the waters within the Gulf of California had turned cold as the waters of the California Current were flowing down the western coast of the Baja Peninsula, around the tip, and into the gulf. Present at this time were the cooler water sharks such as the small gray smoothhound shark, two species of horn sharks, named for sharp spines on their first and second dorsal fins, and an occasional thresher shark. The water was too cold for the semitropical assemblage of fishes present in the gulf during the summer. After looking for members of the assemblage, and seeing none, we spent most of our time mapping the bottom topography at each hammerhead abode with our depth finder. This trip was even less eventful than the fall expedition.

Perhaps what impressed me most during the winter was the behavior of a small species of the manta ray family, called the mobula, leaping like pizza pies tossed into the air by chefs in a pizza parlor. Huge mantas often leapt completely out of the water at the Espírito

Santo Seamount, making a complete revolution in the air before landing on the surface and splashing water everywhere. I have often wondered what could be the function of this athletic "breaching" behavior. Henk, fellow graduate student Giuseppe Notarbartolo di Sciara, and I put on our snorkeling gear and swam over to where all of the mobulas were jumping. Giuseppe was fascinated by manta rays and would eventually describe a new species that inhabited the gulf while earning his doctoral degree at Scripps.

At least half a dozen of these miniature mantas, 3 to 4 feet from wing tip to wing tip, were jumping repeatedly close to shore. Arriving at where they were, we found ourselves immersed in a massive and dense school of tiny larval fishes. These mobulas were invariably landing with a big belly flop, which would produce local displacements of water that would be detected by the lateral lines of the larval fish. The miniature fish dashed toward the center of the school whenever a mobula struck the water, and this compressed the larval fishes into less space. The little manta rays seemed to profit from this response because they captured more of the tiny fish each time they swam through the school with their wide mouths. Alternatively, the displacements may even have been strong enough to stun the larval fish, making it even easier for the mobulas to feed on them.

I had a much better idea of the seasonal changes in weather and fauna of the gulf after the four expeditions in the *Juan de Dios Bátiz* during the period from November 1980 to July 1981. From late fall to early spring, the winds usually blew from the northwest, causing the current to flow out of the gulf along the southwestern coast of Mexico, and then offshore. At this time, cold water from the California Current, flowing southward along the western coast of the Baja Peninsula, at times entered the gulf along the eastern coast. The waters within the gulf were now cold, in the 50s and 60s Fahrenheit, and the tropical inhabitants were absent. From late spring to early fall, the winds usually blew from the southeast, causing currents to flow northwestward along the mainland past Mazatlán into the Gulf of California. Warm water, in the high 70s and low 80s, now moved into the gulf. The arrival of this warm water now permitted the scalloped hammerhead and other assemblage members to enter the gulf. Once within the gulf, the hammerhead sharks seemed to remain at a seamount in the presence of warm oceanic water. The favorite food of the shark, small oceanic fishes and squid, may only be abundant near the seamount at this time.

The mass immigration that we observed at the Gorda Seamount may have been caused either by the cessation of local upwelling of cold water or the arrival of a warm-water mass originating south of the Gulf of California. It is impossible to know which process occurred solely on the basis of the single pairs of satellite images before and after the appearance of the assemblage. For this reason, I sought to relate the coordinated response of many individuals of a single member of the assemblage to a water-mass change based on a series of satellite images of surface temperature. Furthermore, I wanted to record the presence of this assemblage member at the seamount using a technique other than human observation, which was biased by water clarity.

The method I decided to use to answer this question was, once again, ultrasonic telemetry, the technique we had first used in the summer of 1980. On an expedition in July 1986, Don Nelson and I placed beacons with unique ultrasonic signatures on eighteen hammerhead sharks. We traveled to Espíritu Santo aboard the *Robert Gordon Sproul*, a large research vessel operated by Scripps. We did this after placing two tag-detecting devices on the seamount. Their purpose was to record whether the sharks stayed or left as ocean conditions varied. For example, how would the hammerheads respond to upwelling at the seamount? At this time of the year, the strong winds blowing along the coastline caused water to move away from shore, cause nutrient-rich waters below to rise to the surface. The availability of new nutrients at the surface stimulates the growth of phytoplankton. This cool, plankton-rich water could displace the clear and warm water from the eastern Pacific favored by the hammerheads.

Of primary concern was where to put the tag-detecting monitors. The top of the Espíritu Santo Seamount, less than 120 feet from the surface, extends a third of a mile along a northwest-southeast axis. The ridge is 200 yards wide on the northern end and 100 yards wide at the southern end. On the ridge are three pinnacles. The two northern peaks rise to a distance of 60 feet from the surface, while the southern pinnacle comes to within 80 feet of the surface.

We used a crane to lift the panga from of its cradle on the stern of the *Robert Gordon Sproul* and gently lower it onto the glassy surface of the gulf. We then drove the panga over to the northern slope of the northernmost peak of the seamount and lowered an anchor with a line leading to a buoy that floated just below the surface. I then made a scuba dive, during which I attached the tag-detecting monitor, the size of a football, to the line midway between the bottom and the surface. We placed a second mooring with another monitor midway between the middle and southernmost pinnacles of the seamount.

The next step was to find the detection range of the monitors. We attached a surface buoy to the underwater buoy. The surface buoy had a long metal tube run through its center with heavy chain attached to its bottom to keep it upright and a radar reflector on top for detection. We then drove the panga away from the buoy (and the monitor below), stopping for five minutes at each of a series of stations separated by a short distance. At each station, we lowered a beacon to the depth of the monitors and left it long enough to be detected by the monitor under the buoy. We recorded the distance from the buoy to the station from the screen of the radar. We then retrieved each monitor and attached it to a computer aboard the ship to retrieve records of the detection of the tag. We then checked the record for "hits" from the beacon suspended in midwater during those times when we were at the stations, each a little farther from the monitor. The hits were recorded by the monitor on the northern buoy to a distance of 150 yards and 100 yards to the southern monitor. Swimming southwestward over the seamount, a hammerhead would first be detected by the northern monitor and, immediately upon leaving its detection range, would be recorded by the southern monitor.

Halfway through our tests, the line with the beacon transmitter suddenly became tight and then loosened: a fish had taken our transmitter and eaten it. The culprit was probably not a hammerhead, but a wahoo, a torpedo-shaped silvery predator (similar to the Atlantic barracuda) that reached a yard in length. Two or three of these fish usually hovered motionless over the seamount, waiting for an opportunity to dash forward and seize a surprised green jack or pompano. The common name for the species is the Mexican interpretation of the fisherman's exclamation—"Yahooo!"—when the wahoo strikes a lure and dashes off, stripping most of the line from the fisherman's reel.

Once we got the monitors in place, Don Nelson and I began attaching tags to hammerhead sharks within the schools using our pole spears. We started late during the afternoon on the first day, but managed to tag five sharks. We tagged nine sharks during the morning and afternoon of the second day and four more a few days later. Each day, we retrieved the two monitors around noon, acquired the records of tag detections, and replaced the monitors on the moorings. We eagerly looked over the records to see who was and was not at the seamount that day.

The hammerheads exhibited two distinct departure-and-arrival patterns. First, single sharks departed at dusk, moved distances of up to 10 miles into the surrounding oceanic waters (a distance known from the tracking study), and returned to the seamount at dawn the following day. Second, groups of sharks left during daytime and remained away for a few days before returning to the seamount. We had a hunch that these group emigrations and immigrations were in response to changes in the properties of the local water masses.

Although as many as four-fifths of the tagged hammerheads returned to the seamount over the first three days after tagging commenced, the proportion returning dwindled to as little as a fifth of the tagged hammerheads during the next three days. Four separate groups, ranging from two to five tagged individuals, left toward the end of the first period. The water over the seamount was warm during the first three days but was much cooler during the following three days. On the seventh day, there was again warm and clear water at the seamount, and, coinciding with these conditions was the reappearance of two-thirds of the tagged sharks. We had a hunch that the groups of sharks left when cool, upwelled, coastal water displaced the warm oceanic water that had been at the seamount during the first three days of the tagging study.

Again, what had happened was easier to understand when we returned to Scripps and viewed images of sea-surface temperature retrieved from the Satellite Facility. The temperature of the water on these images was indicated by different shades of gray: dark gray indicated the warmest water and light gray denoted the coolest water. A band of dark gray, extended outward from the coast and over the seamount on the satellite images for the first three days. However, a light gray area indicating cool water, became evident within La Paz Bay, west of the seamount, during the fourth day. This mass of cool water grew larger during the next two days, extending around Espíritu Santo Island and over the seamount. The cool-water mass covered the seamount on the day when less than a fifth of the sharks returned. However, the area was dark gray, not light gray on the following day, indicating that warm oceanic water had now replaced the cold coastal water. It was at this time that two-thirds of the tagged sharks returned to the seamount.

During the second three-day period, we were worried that the beacons were not being detected because they were not functioning properly or had fallen off the sharks. We had good reason to worry about this. We had placed the same number of tags on sharks the prior year. Only a few of those tagged previously were detected again at the seamount. That previous year (1985) we had painted bands of yellow, orange, and white on these prototype beacons to help us visually identify the sharks bearing them. We also attached the dart and leader to the underside, not the front of the beacon. I noticed that these beacons wobbled back and forth and the colored bands reflected sunlight as the shark dashed off once it had been tagged. The beacons resembled fishing lures, and it suddenly dawned on me that the sharks might be attempting to eat them.

Sure enough, that is what was happening. I discovered this after I made a free-dive into the school and attached the beacon with the color coding, consisting of three white bands, to a hammerhead. This was, of course, the most reflective beacon that we placed on a shark. Slowly rising to the surface, my attention was drawn to another hammerhead, which dashed out of the school and with its mouth seized the beacon that was floating just above the back of the tagged shark. Upon reaching the surface, I swam over to the panga, in which Don was now sitting. I climbed aboard and told him that I had seen a shark eat one of our beacons. Just as Don said "That's preposterous," I looked over the side of the boat. There, floating at the surface, was the front half of the transmitter with two of the three white bands. I triumphantly grabbed the beacon and held it up in front of Don, pointing to the teeth impressions in the plastic and the metallic barb, which was bent while being yanked from the back of the tagged shark. Despite our frustration, we both laughed loudly. How often did a graduate student have such an emphatic response to a question raised by an incredulous faculty member on his doctoral committee? It was always good to get one up on Don. So during our preparations for the 1986 cruise aboard the *Robert Gordon Sproul*, we painted the transmitters a gray color to match the natural coloration of the shark, and we placed the metallic eye to which the leader was attached at the end of the beacon. These stayed on the sharks, and were used in our study of the effect of cold water upwelling on the residence of hammerheads at the seamount.

In 1986, we had succeeded in relating the movements of individuals of a single species to a local oceanographic process—the upwelling of colder water in the Bay of La Paz. This was no easy feat. We had detected individual hammerheads, carrying electronic beacons, swimming over the seamount by monitoring devices anchored there. Here we were working on a spatial scale of hundreds of meters. The upwelling was described from photographs taken by a satellite from space with a geographic range of hundreds of kilometers (many thousands of meters).

But how does one relate the migratory movements of the many members of this assemblage of fishes to seasonal or annual fluctuations in ocean currents? Ideally, one would like to place sophisticated electronic tags on many individuals of many species while they are at the seamount and track their movements once they leave during late fall on their seasonal migration southward. Do individuals of the same species leave as a group, as the hammerheads did in response to local upwelling? Do the many species occurring at the seamount leave at the same time? Do all of the species travel southward in one massive aggregation? Do they quickly move in a straight line from one place to another, staying at each aggregation site for a while? Do they travel as

I might walk down the walkway to the front door of my house, striding from one stone step to another? Instead of steps, the assemblage may migrate from seamount to seamount or island to island, along a long path leading from temperate to tropical waters. Indeed, there are many islands leading southward from the Baja Peninsula, such as the Revillagigedos Islands, 300 miles south, and Cocos Island, off the coast of Costa Rica, and so forth. These might serve as the migrational stepping-stones. Sophisticated tags now exist that can provide a record of the route taken by an individual over several years and this information can be used to answer all of these questions.

These sophisticated devices are called "archival tags." An archival tag is a microprocessor-based instrument that has sensors to measure either behavioral, physiological, or environmental properties and stores the measurements, or a processed subsample of them, in an electronic memory in the tag until later removal. These measurements can be used to obtain a record of the daily geographic positions of the tagged fish. The ability to infer geographic position on the basis of physical measurements is termed "geolocation." The first generation of geolocating archival tags had to be recovered when the tagged fish was later captured. The instrument could then be connected to a computer for data recovery. The latest-generation tags, "pop-up" archival tags, separate from the fish after a chosen interval, float to the surface, and transmit a series of positions to a satellite, which in turns transmits them to a receiving station situated on land.

Obviously, you wouldn't be able place expensive tags such as these on a large number of individuals of each species. The cost would be prohibitive. Alternatively, one could quantify the degree or association among the species occurring at the seamount. This would be based on repeated visits to the site throughout the year, at which time you would count or estimate the abundance of each species. Two species would be related if many individuals of one occur at the same time as many of another species, or conversely, few individuals of one occur with few of another. You could then select one or two species in a large cluster of related species and track only individuals of these "proxy" species.

Why is it important to track this assemblage of fishes? It consists of many commercially important species—billfish, sharks, tunas, and snappers—which are presently endangered by overfishing. Individuals of these species migrate great distances and do not stay within international boundaries. We need to delineate their migratory paths so that all of the countries along their paths can be persuaded to join together in an effort to conserve them. As I say to my Mexican colleagues, fish are not distributed evenly throughout the ocean like the contents of a bowl of *menudo* (a soup in which fragments of tripe are suspended and that is consumed when one has a hangover), but are aggregated at rocky outcroppings, coral reefs, and seamounts—like the ground meat packed into round balls in a bowl of *sopa del albondigas*. If we find that the assemblage moves from one aggregation site to another, we should strive to create circular preservation zones around the sites and protect the species during their annual migrations.

I vividly recall the passage of this assemblage during the first week of August 1988. I was conducting a submarine survey of the magnetic field north of Seamount Espíritu Santo, because I had started to examine how sharks might use magnetic fields as navigation aids. This was tedious work. We were lowering a magnetometer down to a depth of 200 meters, stopping and holding the device still for half a minute at 25-meter depth stops to measure the intensity of the magnetic field at that depth. We were in position on the next-to-last of twenty stations. Our aim was to map the topography of the magnetic field at the depth at which migrating sharks swam.

We chose this area because the first shark that we tracked had swum along this path a distance of 10 nautical miles, turned around abruptly, and returned along the same path to the seamount on the following morning. It is probably not a coincidence that the shark's path was along the strong gradient between two magnetic lineations, sharply decreasing magnetization to the west and strong uniform magnetization to the east. Seamount Espíritu Santo was situated at the edge of a magnetic plateau with a steep magnetic slope downward to the west. This would make the ideal pathway for the hammerheads to follow during migration.

Louis Zinn, captain of the *Robert Gordon Sproul*, called me by radio from the bridge of the research vessel, which was standing off not more than 200 yards away. He was waiting for us to complete this tedious magnetic survey. He was keenly interested in the natural history of fishes and had worked in the past as the master of a sport fishing boat. He said to me over the radio in an excited tone, "You have to see what I am seeing to believe it." There was a massive procession of whales and fish passing by us and moving in the direction of Seamount Espíritu Santo. The first in this animal parade were twenty or so pilot whales. Then came a crescent-shaped formation of dolphin-fish. Then followed a large school of tuna. Were we privileged to be witnessing the slow migration of this whole assemblage toward its next destination, Espíritu Santo? Could this massive assemblage of whales and fishes be using the magnetic lineation on the seafloor below to guide its procession? Of course, our observation here was anecdotal, a single observation, but it made a great impression on all of us present. It inspired my wish to tag these species and track them on their long-distance migrations in order to determine whether they do move in an assemblage or not and whether they swim from one oceanic stepping-stone to another.

CHAPTER 8

HAMMERHEAD SHARKS
AS OCEAN NAVIGATORS

Three summers of underwater observations of hammerheads from 1979 to 1981 revealed to us the function of schooling. The groups were "harems" of females competing among each other for a central position within the schools, where they were selected by males that later mated with them. The larger females were capable of giving birth to more pups, and hence the reproductive output of the species was maximized. Yet why did the sharks school at the seamount and not elsewhere along the coast or in the open expanse of the surrounding ocean?

During my doctoral studies, I had learned much from my underwater observations about the behavioral activities of the hammerheads during the day but knew little about what they did at nighttime. I had little luck finding sharks while diving at the seamount during the night. The water was dark and forbidding and the beam of my underwater flashlight was narrow and unlikely to illuminate a shark. We needed a better technique to study shark behavior at night. During our second summer expedition to the seamount in 1980, Don Nelson and I attached two ultrasonic beacons to scalloped hammerhead sharks one afternoon and remained at the seamount into the evening listening for the beacons to establish whether or not the sharks remained at the seamount during nighttime. Immediately after the sun set, the signals from the beacons became weaker as the two sharks sped away from the seamount and entered the deep surrounding waters. We left the seamount a couple of hours after sunset, feeling insecure as we were alone in a small boat 10 miles from shore on a moonless night. Yet we managed to find our way back to our "mother" ship, the *Juan de Dios Bátiz*, whose captain was unwilling to stay at the seamount that night and had anchored the boat in a protected bay on Espíritu Santo Island.

A year later, in July of 1981, we visited the seamount in a larger boat, the *Don José*, which was operated by a tourist company called Baja Expeditions. At this time, we tracked three sharks at night, one as far away as 10 miles from the seamount, but we were unable to stay with any of the sharks all night. However, we were surprised to detect each of these sharks swimming among the hammerhead schools during the next day. This was consistent with my seeing the same sharks, identified by their unique color-coded spaghetti tags, on successive days at this time. I then reasoned that the scalloped hammerhead sharks were able to "home"—that is, return to their daytime location after extensive movements in the surrounding waters. They appeared to me to be the homing pigeons of the sea. Salmon, the subject of most homing studies to date in the ocean, migrated extensively in the North Pacific from two to five years before returning to the North American rivers, in which they were born. These fish were certainly skilled navigators, but the scalloped hammerheads might be even better: they made nightly migrations far out into the featureless ocean and were able to find the seamount every morning. The hammerhead seemed a perfect subject for a study of how fishes navigate in the ocean. You could observe more homing movements from the hammerhead shark, which migrated every day, than from the various species of salmon, which migrated only once over a period of several years.

I stayed on at Scripps as a postgraduate researcher after receiving my doctorate in 1982. A year and a half later, I submitted a proposal to the National Science Foundation, asking for a grant to determine how hammerhead sharks were able to find their "home" seamount. I proposed to place more sophisticated transmitters on the sharks with sensors to record elements of their behavior and properties of their environment. I would track the sharks at the surface while acquiring this valuable information. The NSF, which supports basic research throughout the United States, chose to fund my study based on the advice of experts on animal navigation throughout the country. It was on August 16, 1986, that I next tracked a scalloped hammerhead shark during its nightly migration away from the seamount. This time I was able to stay with the shark in my panga throughout the night until it returned to the seamount early the next morning.

That night, there were three of us tracking the shark in my small boat far from shore. The sky was full of stars, and this light gave a gloss to the ink-black water around us. The Baja desert quickly cooled after sunset, while the water remained warm, and dense desert air then started flowing offshore. The winds had gradually grown stronger, creating waves higher than the sides of our boat. Standing at the stern of the boat was the helmsman. His legs were spread apart for better balance as he grasped the aluminum tube extending the length of the outboard throttle. He was having no trouble cutting through the waves, which were moving in the same direction as the boat. It was impossible to see shore, 20 miles distant in the darkness. Barely visible was the white light on the mast of the large research vessel, which was our base of operations. The *Robert Gordon Sproul* had moved into a position between us and the seamount.

We were trying to keep up with a hammerhead shark, swimming away from the seamount at a frantic pace over 300 yards below us. Earlier that day, I had picked out the largest female in the center of a school, swam up to her, and attached an ultrasonic transmitter using a pole spear. She now carried this miniature tracking device, the size of a small flashlight, that emitted short pulses of high-pitched sound exceeding the upper limit of her hearing range. The transmitter emitted short 10-millisecond pulses at a frequency of 38.4 kilohertz (thousand cycles per second); the upper range of the hammerhead's hearing was likely 1.5 kilohertz. The helmsman cut the engine, stopping the boat briefly, so that the tracker could find the direction to the shark. The undergraduate student acting as tracker sat on the front seat of the boat holding a metal staff upright over the side with a handle on top parallel to the surface of the sea. The staff extended underwater 6 feet to a sound-sensing hydrophone. The receptive elements in this underwater microphone were aligned in the direction of the handle so that *bip-bip-bip* from the transmitter was loudest when the handle, visible at the surface above water, was pointing in the direction of the transmitter carried on the shark. Slowly rotating the staff back and forth, the tracker closed her eyes and listened intently for the *bips* to become louder, indicating that the hydrophone was facing the source of the signal. She now opened her eyes and lifted her head to see the handle's direction. Confident with her knowledge of the shark's direction, she held the staff with one hand and used the other to point in the direction toward which the helmsman should proceed.

I sat on the rear seat of the boat staring intently at the bright text displayed on a computer monitor sitting in front of me in a large wooden housing. I lifted up the Plexiglas door and pulled out a movable shelf that held the keyboard. My fingers moved up and down like pistons as I instructed our computer program to convert the stream of *bips* into a record of the shark's behavior and its environment far below. We were using a remote sensing technique, ultrasonic telemetry, to find out what the hammerhead shark was doing in the deep sea, far below and out of our sight. The name of the technique is composed of Greek and Latin words. "Ultra" means "beyond" and "sonic" denotes "sound." These refer to the fact that the frequency of the transmitter's signal exceeded the highest pitch detectable by humans. This is 20 kilohertz, or 20,000 cycles per second. The transmitter's frequency was almost twice as high, 38.4 kilohertz or 38,400 cycles of sound per second. "Tele" means "to send" and "metros" denotes "measurements."

Suddenly, across the screen appeared a series of tiny graphs in bright light. Each consisted of a short, flat line with numbers over the beginning and end of the line with a title over these units of measurements. A recognizable pattern of *bips* sounded over and over again. There were eight repeating *bips*; the time interval between each pair of *bips* indicated the state of one sensor on the transmitter. After each set of eight *bips*, horizontal bars appeared under the lines or axes, of the eight graphs. Some of the graphs provided information on the shark's behavior. For example, the first graph indicated the depth at which the shark was now swimming. Over the beginning of this graph's axis was "0," indicating the surface, and to the right over the end was "350," denoting the maximum depth detectable, in meters. The latest bar displayed on the computer extended almost to the right end of the axis, indicating that the shark was 320 meters below us. The bars on the depth graph grew longer and then shorter as the shark made repeated dives, like the alternating falling and rising of a yo-yo. Yet the bars stayed to the right of the "50," indicating the shark was always swimming deeper than 50 meters. A graph also indicated her heading, or the direction toward which she was swimming. This bar followed the bar indicating depth, and extended far to the right of a "0" and between "270" and "360," indicating that she was swimming in a northwesterly direction. Unlike the bars for depth, the bars for her heading always stayed the same length and indicated that she kept swimming in the same direction. I thought to myself that this was truly amazing! She was now 300 meters below the surface and 700 meters above the bottom and couldn't see the stars or a submarine ridge to use as a reference, yet she kept swimming in a perfectly straight line in a fashion similar to the way one might drive a car down a highway by looking at the center line.

Other graphs indicated information about the deep-sea environment in which the shark swam. The next bar was halfway between "10" and "30," indicating the water around the shark was 20°C or 68°F. The next two graphs measured the amount of irradiance present undersea. I use the word "irradiance" instead of "light" to distinguish between the wavelengths of electromagnetic energy visible to humans and those visible to sharks. "Light" is used to refer to the range of wavelengths to which humans are sensitive. Sharks are most sensitive to blue-green, and this makes sense, because sunlight of these wavelengths transmits farthest underwater. The red light, to which humans and other terrestrial animals are sensitive, is rapidly absorbed underwater. Marine animals live in a blue-green world. The sun is not the only source of irradiance in the ocean. Euphausids (shrimplike animals), squids, and fishes that live in the deep sea during the day and migrate near the surface during the night often have photophores covering their bodies that emit irradiance in the blue-green wavelengths. We measured irradiance over two slightly different spectra or ranges of wavelengths. One range matched the spectral response of light-sensing chemical compounds in the eye (called pigments) that were sensitive to high levels of electromagnetic energy and the other to pigments sensitive to low levels. Humans, like sharks, have different pigments that work in bright or dim light. The switchover from one to another pigment is evident in a brief period of blindness as our eyes adapt to dim light after turning off a bright light in our bedroom.

I now looked back at the computer monitor. The bars that appeared under these two graphs, to the right of the others, were very small, indicating little "shark light," or irradiance, was present where the shark now swam. This confirmed my hypothesis that hammerheads do not use the stars or even the sun for guidance as do terrestrial organisms such as birds. There was simply no light to guide the shark: the bars were absent, indicating that there was almost no available light where she was swimming. Yet she was swimming away from the seamount in a perfectly straight line.

It was exciting to learn what the hammerhead was doing far below us at night. Sure, a day earlier we had been able to keep up with the schools and watch the sharks within them. As the sun dropped to the horizon and there was less light underwater, I had observed the school slowly break up. First, members left the tight formation, slowly circled alone for several seconds, and returned to the school. Eventually, they no longer returned and began to swim rapidly alone or in small groups of two or three. It was hard to keep up with these fast-moving sharks. The sharks would even swim faster when swimming away from the seamount. They would quickly outdistance me at this time, swimming at their customary cruising rate of one meter per second, a speed equivalent to a fast-paced walk. Sharks resemble humans in having two modes of locomotion—sustained swimming at a rate of 1 meter per second and sudden bursts of swimming of rates greater than 4 meters per second, a dichotomy similar to our walking and running. I finally had to give up following them because it became too dark to see them underwater.

We were now able to stay on top of the shark because she had finally slowed down. There was little change in the loudness of the transmitter's signal as the tracker rotated her staff in a complete circle, but the signal became much louder when she tilted the staff and hydrophone downward into the depths of the ocean. The shark now was 300 meters deep. She stayed at this depth and moved very slowly, but now changed direction often. I had a hunch: was she changing direction after seeing a flash of bioluminescence from a squid and dashing off to capture it? I looked over at the two irradiance graphs on the monitor, and, sure enough, both bars had lengthened. She had suddenly been illuminated by blue-green irradiance. Where was this light coming from so far from the surface? No light could penetrate to that depth when the night was pitch black, with no moon overhead. The flashes of light must come from light organs of a deepwater squid. After all, we had collected several species of squid covered with photophores from the stomachs of hammerhead sharks caught near the seamount. On this particular night, our shark seemed to have two distinct types of travel, a fast and highly directional mode that consisted of "yo-yo" diving oscillations and a slow, nondirectional swimming mode, at which time she stayed in one place and the same depth. It sure seemed like she had located an aggregation of squid and was gorging herself on them.

Just after midnight, she suddenly swam fast again, but now in the opposite direction, back toward the seamount. There in front of us in the distance was the dim mast light of the research vessel. The helmsman looked on the monitor of our radar set mounted on a post attached to the rear seat. A green trace of light moved in a wide circle around the screen, illuminating a large target five miles to the southeast of us toward the seamount. He exclaimed, "Wow, we must be ten miles from the seamount now, if the research vessel is halfway between us and Espíritu Santo." The captain of the research vessel was, in fact, looking on his radar for the direction and distance to a huge helium-filled balloon with a radar reflector that we had moored over the seamount earlier that week. And now the shark was heading directly back to the seamount and might even pass under the research vessel on her way home. She seemed to be returning along her outward path. This struck us as a formidable task. The tidal currents had surely changed direction since we tracked her away from the seamount, and they should deflect her from her course. But they didn't do that. It was now difficult to keep up with her as she sped home because the helmsman was driving the boat in the direction opposite to that of the waves. *Bam-bam-bam* . . . The waves hit our bow over and over again, splashing water over the boat. It was also a struggle to operate the computer equipment while keeping it dry at the same time. However, we managed to stay behind the hammerhead until it arrived at the seamount at 6 A.M. as the sun rose above the horizon.

This was the first of many scalloped hammerhead sharks that we tracked during their nightly excursions away from Espíritu Santo over a four-year period from 1986 to 1989. How directional were the movements of the hammerheads? The first shark that we tracked swam with great directionality 10 miles from the seamount before changing direction and returning along the path taken during its outward movement. Not only was her path straight over ground; she swam with a constant heading while moving away and returning to the seamount. There was no lessening of orientation with increasing distance from the seamount as I had anticipated. This really surprised me. These sharks were swimming as if they were keeping on a predefined course through the darkness of the ocean.

During the following year, a transmitter came loose from a shark, one that we had tracked for five days, and, to our great chagrin, it floated to the surface. The transmitter, shed at 10 P.M. in the evening, was tracked while floating at the surface until 6 A.M. the following morning. During this period, the tracker was shocked by what seemed to be the unexpected behavior of the hammerhead. She moved little while staying at the sea surface while her headings varied constantly. It wasn't until dawn that the tracker realized what had happened, when the helmsman spotted the transmitter floating at the surface.

What initially struck us as a real setback was later considered a stroke of luck when we plotted the track of the transmitter drifting at the surface. The transmitter was initially transported slowly in the current toward the southwest, then sat in one place for several hours, and finally drifted away from the seamount toward the southeast as the current picked up speed with the change in tides. The transmitter, drifting at the surface, traveled less than a fourth of the distance that our first shark traveled during the same time period. Furthermore, the transmitter, detached from the shark and floating at the surface, spun around in all directions. These were not at all like the headings of the first shark tracked, whose swimming direction was constant in a north-westerly direction. The erratic orientation of that transmitter at the surface made us appreciate the ability of the shark to orient itself in the darkness of night, far away from any potential directional reference at the surface or bottom.

A pattern emerged as we tracked other hammerhead sharks at the Espíritu Santo Seamount. We followed one female to and from the seamount seven times, another back and forth six times. A map showing the homing movements of the three sharks was made using a drafting program in the dry laboratory aboard our research vessel. We plotted the tracks using the bearings and ranges (or distances) from the panga to the vessel and similar bearings and ranges from the research vessel to the balloon over the seamount. The scientific party aboard the ship met to discuss the tracks. It was apparent to all of us that the tracks of the hammerheads on the map radiated from the seamount like the spokes from the hub of a wagon wheel. All three of the sharks, including the female tracked earlier, took the same path leading away from the seamount toward the north. We named this Route 1. There were five more routes leading away from the seamount. These "roads" were taken more than once by the same shark and by different sharks on separate evenings. The sharks usually returned to the seamount via the same path along which they had departed.

Route 6 differed from the others. The main route proceeded in an easterly direction until it branched, with one branch extending north-ward and the other branch continuing westward. The hammerhead shark first took this route in a easterly direction, turned left to go northward, reversed its direction and retraced its path southward, continued farther westward, and then turned around and came back to the seamount. How extraordinary it was for this shark to retrace her steps back to the seamount when the currents should have deflected her away from her prior trajectory. As demonstrated by the floating transmitter, currents could change direction over the period of a track. She would have to be very smart to anticipate those tidal changes and compensate for them on her way back. Would it not be much simpler for her to swim above some conspicuous feature on the floor of the ocean? The commuter, of course, steers the car along one lane of an asphalt road split from the other lane by a white line. The hammerhead shark could obviously not see the bottom, but near the surface it could perceive a pattern of the local magnetic field.

A basic challenge for any fish living in the ocean is to know where it is. Have you ever wondered how a school of fish remains in one place over a reef with its members feeding on plankton floating past in the current? The school may be 10 fathoms above the bottom and a similar distance from the surface. The eyes of the school members are situated on the front of their heads to see the passing plankton, not on the top or bottom of their heads, from where a feature on the surface or bottom could be detected and used as a reference to stay in one place. The challenge is even greater for a migratory fish such as a scalloped hammerhead shark that moves from one place to another. What does a species use as a reference to guide its movements—in the case of the hammerhead shark, when migrating from its "home" seamount to feeding grounds over other seamounts? The patterns of magnetization in the seafloor would be of great benefit as landmarks to guide the movements of any wide-ranging oceanic animal—these fixed patterns exist at the surface far from the bottom. They would be a far better positional reference than the alteration of the direction of water currents as they flow over prominent features in the bottom topography, because these changes in current flow do not likely reach the surface.

To understand the potential value of this feature, we need to know more about the magnetic field of the earth. The total magnetic field measured at the earth's surface is the sum of two magnetic sources, the earth's inner core and its outer crust. Circulation of electrically conductive molten metals in the earth's core creates the dipolar (two-pole) nature of earth's main magnetic field; hence, the north and south poles. This is a very strong field. The needle of a compass points to the north pole—it is oriented by the earth's main magnetic field.

Magnetic minerals (oxides of iron and titanium) in the earth's crust produce minute distortions, or "anomalies," in this uniformly dipolar global magnetic field. These materials form patterns. Two are particularly common on the seafloor. Magnetic dipoles, two poles less than a mile or two apart (like the earth's poles but much closer together), are often present at underwater mountains, or seamounts, due to their volcanic origin. These deposits in seamounts were formed during past volcanic eruptions from molten lava that contained tiny magnetic particles, which aligned, like microscopic compass needles, along the north-south axis of the earth's dipolar field. Over geologic time, the polarity of the earth's field stays constant for a long period, reverses quickly, and remains with the opposite polarity for a similar period. A second eruption, forming another part of the volcano, may occur when the earth's polarity is reversed—and then the magnetic particles align in the opposite direction. Particles of one polarity thus exist in one part of the seamount and the other polarity in another part. Their additive and subtractive natures create positive and negative anomalies that together create a small-scale magnetic dipole. Lava may also flow out onto the seafloor in a single direction and cool with the particles aligning to the polarity of the earth's field. A valley or a ridge in magnetic intensity is created depending on whether the polarity of the particles subtracts from or adds to the current dipole of the earth. These valleys and ridges might serve as "roads," along which the hammerhead sharks could move to and from the seamount and surrounding waters.

A second pattern of magnetization, common to all ocean basins, is a large-scale, pin-striped pattern of alternating bands of strong and weak magnetization (50 to 100 miles across) leading in a north-south direction. Molten lava laden with particles of magnetite continually rises to the surface of the earth at the center of the oceans and spreads outward to either side to form new ocean floor. As the rock cools, the magnetite particles are permanently embedded in the rock with the polarity of the earth at that time in geological history. After several hundred thousand years of spreading, the earth's polarity switches over a few thousand years, and now the particles in the spreading crust become embedded with the opposite polarity to those in the earlier deposited crust. Steep gradients (or changes of intensity with distance) exist between these bands, in which exist magnetic particles with their dipole movements either toward or opposite to the present direction of the earth's dipole axis. The boundaries between the steep and weaker adjacent gradients could serve as another geographic reference—pathways along which marine animals might migrate between temperate and tropical environments.

There was some evidence relating the movements of marine animals to the patterns of seafloor magnetization available when I began to do my research into shark navigation in 1984. Strandings of whales frequently occurred at certain points along the northeastern coast of North America and the coast of Great Britain. These stranding centers were located where valleys in the local geomagnetic relief intersected the coastline. If the whales were using the bands of strong and weak magnetization to guide their movements, then the coast would have blocked the forward progress of the whales.

It became a priority for me to map both the bottom and the magnetic field surrounding the Espíritu Santo Seamount. The paths of the sharks could then be compared to the orientation of ridges and valleys in the seafloor topography as well as maxima and minima in the local magnetic field. I would have to map the bottom and the magnetic field from the seamount to the most distant nighttime destinations of the sharks. I conducted a combined bathymetry and magnetometer survey at the seamount with the *Robert Gordon Sproul* during July 1988. For four days, we circled the seamount in a series of concentric circles, each one a mile farther from the peak of the seamount. The fathometer and magnetometer produced an oscillating trace of bottom depth and magnetic intensity on the continuously moving strip of paper. We periodically recorded measurements directly off the ship's fathometer and magnetometer. This was twenty-four-hour work, with three shifts of Mexican and U.S. scientists working eight hours per shift. We made concentric, rather than parallel, survey lines because a more regular survey path resulted from steering the vessel with a constant rudder angle, while using the ship's radar to maintain a constant distance between the ship and the helium-filled balloon and its radar reflector suspended above the peak seamount.

The magnetic survey was complicated by the fact that the intensity of the magnetic field in any one place varies over the day as well over a season. It would take us four days to collect measurements along the twelve circles around the seamount. We needed to compensate for the variation in magnetic intensity for that particular time of the day when each measurement was made. We transported a second magnetometer in our panga to the Isla Espíritu Santo, the closest island and 12 miles distant from the seamount. We then carried the bulky instrument and two very heavy lead-acid batteries to a place on the island desert, where its caretakers could look out on the water and see the research ship slowly moving around and around in ever larger circles. A tent was pitched next to the magnetometer. Here members of the island crews could prepare meals and sleep when not attending to the magnetometer. An intensity was measured using the magnetometer on the island every five minutes, when a measurement was made in the survey area.

This was really a tough and unpopular job. During the day, the summer heat made everyone on the island feel like they were being cooked alive in an oven, but in the evening there were dense clouds of *jejenes* (pronounced hay-HAY-nes in Spanish), insects that swarmed around us and drew our blood like mosquitoes. One twenty-four-hour period was all anyone could tolerate. For this reason, the crews rotated on the island. This also was tricky business when transporting crews back and forth to the island, because those in the boat departing from the island, in order to be picked up, had to reach a designated pickup spot at exactly that time when the research vessel passed by the island on one of its wide circular excursions.

The ship-based measurements would eventually be "normalized." We either subtracted or added the small difference between the field strength at the stationary site and a single intensity measured on the island at 8 P.M. Why use a reference measured at this particular time? It was the time when the female shark, mentioned earlier, exhibited her most highly directional point-to-point movements and teleme-tered swimming directions. The measurements of bottom depth and magnetic intensity and corresponding geographic coordinates were then entered into a mapping program to produce 3-D bathymetric and geomagnetic contour maps of the seamount area.

We all stood around the large table admiring our maps of the Gulf of California around the seamount. The seamount was the highest peak in a range of underwater mountains that extended in a north-ward direction. Eight miles to the east, and separated by a deep valley, was a broad underwater plateau. One of the sharks swam over this valley during four different nights. She swam rapidly and directionally when over the valley but slowed down and moved aimlessly back and forth when swimming over the plateau.

The map of the magnetic field shed light on the female shark's nighttime commute from the seamount. Her path away from the seamount was located directly on the boundary between a band of weak and strong magnetization on the seafloor. In fact, the other two sharks had also taken this trail during nightly excursions. However, the other two sharks often also swam to the underwater plateau east of the seamount. Their tracks fit over a series of magnetic valleys and ridges that radiated outward from the seamount like spokes from the hub of a wagon wheel. Perhaps these valleys and ridges might be ancient lava flows, which hardened with their magnetic particles ori-ented toward, and opposite to, the present polarity of the earth's field.

Each of the hammerhead sharks, composed of ionic fluids, induced an electromagnetic field as it swam through the earth's magnetic field. The sharks can be considered a biological version of a copper wire, which is passed through the field of a bar magnet. The intensity of this induced electrical field depended on the speed and direction of the movement of the conductor (the ionic fluids in the shark's body) rela-tive to the local magnetic field. The hammerheads have electric recep-tors on the underside of their snout capable of detecting these induced fields. Could these sharks be comparing the minute differ-ences in geomagnetic intensity perceived by the receptors located on the left side of their heads to the intensity perceived on the right? The lateral elongation of the head would amplify any intensity gradient because the difference between two intensities would be greater at more widely separated points. Furthermore, there are more electric receptors on either side of a wide head, and these should provide the shark with a greater ability to discriminate between the two different intensities.

I refer to this style of orientation as geomagnetic "topotaxis." It is the ability to orient to local maxima or minima in geomagnetic intensity. The prefix "topo-" refers to the relationship between movements and topography. In the dictionary, "topo-" is defined as "the configuration of a surface including its relief and the position of its natural features." The suffix "-taxis," a scientific word, indicates that an animal is attracted to a property in the environment. In this case, the shark actively tracks magnetic ridges and valleys, features of relief in a surface of geomagnetic field intensities.

It is essential to distinguish topotaxis from a compasslike sense, which is the ability to maintain a constant heading using directional references such as the sun, the moon, the stars, or the earth's dipolar main field. These styles of orientation may best be distinguished from each other with an analogy to the quite different methods by which humans navigate large airplanes and helicopters. The airplane pilot navigates between two widely separated geographical points by steering in a direction relative to the northward orientation of the compass magnet, utilizing knowledge of the difference between the former direction and that of the destination. The pilot returns by flying in a complementary direction. If the wind speed is strong and perpendicular to the plane's course, their airplane is deflected from the direction of the destination point. The pilot compensates for this deflection by changing course based on the knowledge of the wind speed and direction. The resulting flight path is often slightly curved. Similar paths are expected from an animal with a compass sense.

A helicopter, on the other hand, is often flown in relation to local features and, therefore, navigated differently. The helicopter pilot visually follows a road, valley, or ridge, resulting in a path that can be sinuous. The path of the helicopter depends on the path of the reference feature. Straight roads give straight flight paths; winding roads give winding paths. This is the most simple, or parsimonious, explanation for how the sharks could swim away from the seamount and return along the same path. The hammerheads are likely detecting a "road" not with their eyes as a human would, but with their electric receptors. The shark's navigational connection with its world is not a visual one like humans', but a geomagnetic one. What a powerful capability this would be in the apparently featureless expanses of the ocean. It is critical to be able to swim in one place at the seamount without always having to see the bottom or to travel from home to one's feeding grounds without having to see the bottom. Vision has limited value when the water is turbid or dark.

The paths of the sharks on our map of the local magnetic field were at times parallel to the magnetic contours (lines indicating a constant magnetic intensity), but at other times crossed them. This inconsistency bothered me. I wanted some better evidence that the sharks were orienting to the magnetic ridges and valleys leading from the seamount. This relationship became more apparent when I looked at the record of the magnetometer. As the ship traveled around the seamount, the magnetic intensity was recorded as an oscillating trace of black ink on a continuous roll of graph paper. Every five minutes, we placed a number on the record to indicate our current position.

I later cut the long paper record of the magnetometer into ten segments, each showing the trace of magnetic intensity recorded during one circular passage of the vessel around the seamount. Each segment was one nautical mile farther from the seamount and longer than the one a mile closer. I then attached the shortest segment, which showed the trace of magnetic intensity recorded as the vessel circled nearest to the seamount (only a mile away) to the bottom of a wall in my laboratory. I next placed the segment recorded during the circular passage of the vessel one nautical mile farther from the seamount immediately above the first, and so forth until the segments covered the wall and reached the ceiling. I now stood back and looked at the collection of magnetometer records. You could locate the same valleys and ridges leading away from the seamount on several segments, each farther away from the seamount. I then drew a unique symbol on each segment where a particular shark passed over the path of the magnetometer on its migration to and from the seamount. I now looked up and down the wall on the various segments for the unique symbol indicating the path of each homing shark. I was stunned at what I found. The outward migration of each shark was confined to the same valley or ridge, apparent on each successive segment of the record farther away from the seamount. Furthermore, its return migration was also confined to the same valley or ridge. And where were the positions of the large female hammerhead that we tracked first? I looked down at the small segment recorded as the vessel circled nearest to the seamount— it was attached to the base of the wall. The black line, indicating magnetic intensity, rose continuously until it abruptly leveled off at the edge of the magnetic plateau, the site of the seamount. This edge was evident above on the segment next farthest from the seamount, and so forth. Solid circles indicating the path of the large female moving to and from the seamount were apparent above each other on this unique point in the trace. Here was Route 1, the path taken by the large female and other hammerheads during their nightly travels away from the seamount. This was even stronger evidence that the sharks were using geomagnetic topotaxis to guide their movements.

What is necessary to prove conclusively that hammerhead sharks sense the magnetism of the earth? Relating the tracks of the sharks to the magnetic valleys and ridges is not conclusive proof because some other guiding feature unknown to us might exist at the same place. You must perform an experiment: alter the magnetic pattern and look for a predicted change in the shark's behavior. How does one reverse the magnetic dipole at a seamount or displace a magnetic ridge leading from it? I proposed several years ago to encircle the seamount three to nine times with copper wire in order to make a huge electronic coil that, when energized by two lead-acid truck batteries, would reverse the polarity of the whole seamount. This was not as impractical an idea as it might seem. Engineers degauss, or neutralize, the magnetic fields of massive battleships and aircraft carriers, which are constructed of many ferrimagnetic metals, so that the ships do not detonate magnetic mines during war at sea. Why not reverse the dipole field at a seamount?

During the summer of 1986, I tagged eighteen hammerheads with coded ultrasonic beacons over the north end of the seamount. We recorded whether they stayed on the north or south side of the seamount with a tag-detecting instrument moored at either side of the seamount. The monitoring devices, each of which could detect a tag at a distance of 150 yards, were separated by just over twice that distance. Thus the sharks were expected to spend the same amount of time within the detection range of each monitor. However, this was not so! The hammerheads were highly selective about where they stayed on the seamount. The sharks were detected by the monitor on the north side of the seamount all day simultaneously moving in and out of its range as expected of sharks moving within a school. However, only once did one of the tagged sharks stray over to the south side of the seamount to be detected by the second monitor. This is certainly odd for sharks that at times swam 10 nautical miles from the seamount at night to feed in the surrounding waters. There was something special about the north side. My hope someday is to temporarily reverse the dipole field so that the south side of the seamount now has the prior polarity of the north side, and vice versa. Would the hammerheads move in a school over to the south side of the seamount, now that its polarity was familiar? The two tag-detecting monitors moored at either end of the underwater ridge would provide the answer to this question.

In 1984, before the study began, I was running while kicking a soccer ball on the sandy beach in front of Scripps during lunchtime, as was my custom every day. Wolfgang Berger, a tall, gangly geophysicist who taught part of the marine geology course to graduate students at Scripps, came over to me. He asked me what I was studying now that I had gotten my doctorate. I told him that I wanted to know how schools of hammerheads could find a seamount. He laughed, and said jokingly, "It's easy to find a seamount." The hammerhead would find a seamount like any good geophysicist. This scientist would use a gradiometer, consisting of two magnetometers spaced slightly apart from each other. The instrument would be towed around the area, searching for the strong gradient associated with the seamount. Everyone knew that strong dipole fields were associated with seamounts. It wasn't until four years later, after expending much of my energy tracking hammerhead sharks and even more energy conducting a magnetic survey, that I arrived at the same conclusion.

Why form schools at the seamount? It is a landmark at the center of the hammerhead's home range with many magnetic pathways radiating outward to various feeding grounds in the expanse of deep water surrounding the seamount. It is convenient for all of the sharks to remain there, schooling in one place during their resting period of the day so that they can carry out their social activities before separating from the schools and migrating outward along these familiar pathways to their nightly foraging grounds. In a sense, the hammerhead performs the reverse of the human commute. Modern commuters work during the daytime in the city and sleep during nights in the suburbs—hammerheads relax in the city during the daytime and forage in the suburbs during the night.

CHAPTER 9

IN QUEST OF THE
WHITE SHARK

To an ethologist such as myself, interested in learning about the behavior of animals in their own environments, finding out the true nature of the man-eater, or white, shark was an irresistible challenge. This became an all-consuming ambition of mine after I completed my doctoral studies on hammerhead sharks at the Scripps Institution of Oceanography in 1982. First, though, I had to learn more about the life cycle of the white shark in order to find where I could best study the shark in its own habitat. For this reason, I searched through the scientific literature for all of the published records of white sharks caught off the western coast of North America. I then visited marine laboratories, oceanariums, and natural history museums in California and examined their unpublished records. These records, going as far back as 1955, contained the date and location of the capture of each shark, its size and weight, and, usually, its stomach contents.

The following picture emerged from these records of the white shark's life cycle. Pregnant females visit the islands off Southern California during summer and fall to give birth. These females, more than a dozen years old, exceed 15 feet in length and can weigh over a ton. They are attracted to their favorite prey, northern elephant seals, which live in large colonies on these islands. Born alive and fully independent, five to ten baby sharks from a female swim toward shore along one of the submarine canyons, which lead from the continental shelf surrounding these islands to the coast of California. These newly born white sharks, averaging 5 feet in length, probably travel together within a school. Evidence of this is contained in the field notes recorded for a white shark specimen kept in the fish museum of Scripps. In October of 1955, Jimmy Stewart, the dive master at Scripps, was surprised while teaching his dive class to see a small school of white sharks swimming near the Scripps Pier. He and his students set out a long line and caught eight juvenile white sharks within two weeks—all were between 5 and 6 feet long (three exactly 5 feet 4 inches long) and less than a year old—their similar size indicated that they likely came from the same mother. The Scripps Pier is only a quarter mile away from the La Jolla Submarine Canyon, which is inshore of San Clemente Island. Juvenile white sharks are also frequently caught near the Ventura Submarine Canyon, which is located directly east of the northern Channel Islands—Anacapa, Santa Cruz, Santa Rosa, and San Miguel. Several years ago, a television crew in a helicopter filmed a school of baby white sharks swimming near the surface in Santa Monica Bay near the Santa Monica Submarine Canyon, which is directly inshore of the southern Channel Islands— Santa Barbara and San Nicolas.

These baby white sharks have pointed teeth for seizing and swallowing prey—unlike the triangular and serrated teeth of adults used for cutting off pieces of meat from pinnipeds (the name given to the taxonomic group that includes sea lions and seals)—and chase down and feed on fish such as the cabezon. These fish live near the bottom over rocky reefs. After spending two years developing in Southern California waters, the juveniles move northward along the coast during either summer or fall. By three years of age and close to 9 feet long, the sharks begin to feed on seals and sea lions. These two types of prey are easy to distinguish from each other. Seals have small foreflippers and highly developed hindflippers, which are used to propel them underwater and are of limited value on land. Sea lions, to the contrary, have well-developed foreflippers, which are used to swim underwater and enable them to walk on land with a shuffling motion. Seals are restricted to the beaches and flat areas of coastline because they are poor climbers out of the water—the stout, sausage-shaped seals slowly crawl ashore with an inchworm motion. Sea lions, on the other hand, cluster in great numbers high up among the rocks along the entire coastline. They are able to climb up on the rocks using their foreflippers. Adult white sharks spend the fall months of each year in Central and Northern California near colonies of seals and sea lions. White sharks had been caught near the colonies of seals and sea lions situated at the Point Reyes Headlands, a peninsula jutting out into the Gulf of the Farallones north of San Francisco, at Año Nuevo Island, at the northern edge of Monterey Bay, and at the Farallon Islands, 30 miles west of the mouth of San Francisco Bay.

Critical to understanding the distribution of white sharks is an understanding of certain aspects of the life cycle of the northern elephant seal, its favorite prey. Females at the elephant seal colonies give birth and suckle the pups to provide them with nourishment from December through early February. The pups begin entering the nearshore water, learning to swim and dive, in mid-February, and by late April they have all migrated away from the colony. Although we don't know exactly where the juveniles go, it is likely that they travel considerable distances, because the members of other age classes do. Adult females meander over a broad expanse in the middle of the eastern North Pacific; males forage along the continental margin of Oregon and Washington to the upper reaches of the Gulf of Alaska and eastern Aleutian Islands. While at sea, the juveniles appear to be less vulnerable to predation than when they rise to the surface to breathe near the colony—they spend 90 percent of their time at sea submerged and almost all of their dives are greater than 200 yards deep. White sharks have well-developed eyes and associated muscles that help them focus on and pick out silhouettes of objects on the surface. It is when the juveniles converge on the large seal colonies at the Point Reyes Headlands, Año Nuevo, and the Farallon Islands during September through November, after their first trip to sea, that these naïve juvenile seals are at the greatest risk of being captured and eaten by a white shark.

It was thus with great interest that I listened in the spring of 1983 when David Ainley, a biologist at the Point Reyes Bird Observatory, told me about the white sharks at the Farallon Islands. I was making the rounds between the various marine institutions in Central California in order to examine any records that they had collected on white sharks captured locally. I wanted to meet with David and visit the Point Reyes Bird Observatory. It was a cool, foggy day, so common along the coast of Central California. He was leaning back against a railing on the gray, wooden deck in front of the observatory. This was a tiny research station, consisting of a cluster of rustic, wooden buildings, nestled in the lush green forest next to Bolinas Lagoon north of San Francisco. Everyone referred to the observatory by its initials: PRBO.

David stood there, dark disheveled hair, bushy black mustache, staring at me with an intense gaze. He spoke for a while, then paused for quite a long time, which made me a bit uncomfortable (he surely expected me to reply, but I wasn't ready yet), before he continued to speak again. It surprised me that he spoke with such economy of words, which was in stark contrast to the many papers by him in the scientific literature. Dave was truly a Renaissance man, writing about the biology of species ranging from sea lions to birds and, of course, to white sharks. I had read with keen interest an article that he and the other biologists of PRBO had published two years earlier about white sharks feeding on seals at the Farallon Islands.

He told me that PRBO had maintained a year-round research station since 1968 on Southeast Farallon, the eastern of two islands that make up the South Farallons, (the other island is called West End) situated 30 miles west of San Francisco Bay. Southeast Farallon is the larger island, one-third of a mile across from east to west and a quarter mile from north to south. Rising 340 feet above sea level on the northern side of Southeast Farallon is Lighthouse Hill, a pyramid-shaped hill with a tower containing a rotating light that casts a bright beam into the foggy darkness. The rotating light and loud nasal sound of the foghorn warn local mariners of the peril of running aground on the rocky shores of the island. The PRBO facilities are contained in a Victorian-style house situated on a flat terrace on the southern side of the island and facing southward toward a large crescent-shaped bay. To the southwest is the smaller West End, which is separated from its neighbor by a narrow surge channel, only a dozen yards across at its widest point. West End is also dominated by a large hill, Maintop, located in the center of the island, and a tall ridge to the west of it and along the coast that is oriented toward the northwest. The two islands are collectively half a mile across.

Two biologists, plus up to six additional volunteers, stayed at the PRBO station on a rotating, but continuous schedule. Every day, weather permitting, these people walked around the island, conducting a census of the birds on the island and reporting what happened in the waters close to shore. During the fall season, they had observed white sharks feeding on seals close to shore. Most of the attacks occurred in front of the field station in the large bay off a small peninsula on which elephant seals congregated. The number of attacks observed had increased to seven per year during the last nine years, and the frequency of attacks appeared to be increasing as more and more seals colonized the island.

David said that PRBO wanted to work with me to tag and track white sharks at the island. He urged me to quickly put together a proposal to do this work and send it to the National Park Service, which had indicated a willingness to fund this type of study at the South Farallon Islands. There were things I needed to know about working on the island. A permit was required from the United States Fish and Wildlife Service; they were custodians of the island, which was a nature reserve. One assistant and I could stay at the field station, the two-story house in which the biologists and volunteers lived. Supplies were delivered to the island every weekend aboard the pleasure boats of members of the Oceanic Society, a San Francisco group devoted to ocean recreation. On the eastern side of the island, there was a large crane powered by a generator housed in one of the buildings. The crane could lift my Mexican panga out of the water and set it onto a concrete platform 25 feet above the water. If my boat were too heavy to lift, the panga could be tied to one of two buoys moored close to the island. I could then go back and forth from skiff to shore using the PRBO inflatable. I should weigh the boat right away to ascertain whether it weighed less than the crane's 5,000-pound lifting capacity. The crane was fastened to a concrete platform that had a concrete block with a metal eye for lifting, which was hoisted periodically to test the crane's weight capacity. I should plan on working there during September and October, when juvenile elephant seals, the favorite prey of the sharks, were abundant. These attracted the sharks to the island. Furthermore, during fall the winds were weak and the water often calm, ideal conditions to operate a small boat in the waters surrounding the island.

At 5:30 A.M. the next day, I boarded a 36-foot sailboat, kept at Pier 39 in downtown San Francisco, and by 6 A.M. departed on the five-hour journey to the Farallon Islands. The plan was to motor out of the bay and then sail the rest of the way to the island. The water was perfectly flat as we moved through San Francisco Bay. I looked up at the massive bridge, over a mile long, suspended between two towers, leading from the city to its northern suburbs of Marin and Sonoma. The Golden Gate Bridge, illuminated by the rising sun, was a rusty red color from its coating of anticorrosion paint.

Once underneath the bridge, the boat began wildly rocking back and forth in atypically rough seas. We had inadvertently left the shipping channel and were over the shallow "Potato Patch." The current had quickened as the large aqueous contents of the bay passed over the shallows and moderate winds were blowing from the opposite direction, and that made for high wave crests and little space between them. In fact, some of the waves were breaking, and the sailboat might capsize if it was caught on one of these waves. The situation was dangerous, and to my surprise the captain asked me to steer the boat. I actually enjoyed taking the helm in such perilous conditions, confident in the sailing experience I'd gained by living four years on a sailboat while a graduate student at the University of Miami. We escaped from the Potato Patch as quickly as we strayed into it. Once we were halfway to the islands, we raised the main sail and started sailing toward the island.

After another two hours, the island gradually appeared in the distance out of a thick blanket of white fog. The color of the water around us was steel gray, unlike the light blue water of the Gulf of California. As we drew closer, we could see the island better. It was composed mostly of brown rock, covered here and there with the white guano (feces) of the many seabirds that made it home. In the center of the

1) A white shark biting a piece of bait. This behavior, though rare in the daily activities of a shark, is the image of the species one most often sees on television. (VALERIE TAYLOR)

2) The majority of species of sharks were considered extremely dangerous when I first entered the water to study them. So for safety I made my early observations from within a cage. (ART MYRBERG, JR.)

In my killer whale costume (3) and swimming in the lemon shark pen with the costume (4). Notice the white *H* on my wet suit, resembling the coloration on the killer whale's belly, and my wooden dorsal fin, painted black to resemble the dorsal fin of a killer whale. The sharks reacted to my approach by performing an aggressive display. (GERRY KLAY)

5) I attracted blue sharks to a container that dispensed ground-up fish. The sharks swam faster and began to bite the container as more individuals were attracted to the boat. Animal behaviorists call the heightened activity of individuals in the presence of others "social facilitation"; the public refers to the same behavior seen in sharks as a feeding frenzy. (JEREMIAH SULLIVAN)

6) My colleagues and I made six ten-day cruises aboard the *Juan de Dios Batiz* to study schools of hammerhead sharks in the southern portion of the Gulf of California. The picturesque fishing trawler was operated by CICIMAR, a marine laboratory in La Paz, Mexico. (TED RULISON)

7) Three hammerhead sharks swimming over the Espiritu Santo Seamount, which rises from a depth of over half a mile to 55 feet below the surface. (TED RULISON)

8) Donald Nelson, a member of my doctoral committee, helped me mark hammerhead sharks with spaghetti tags at the Espiritu Santo Seamount. Here Don is about to tag a shark at the edge of a hammerhead school. (FLIP NICKLIN)

9) It was often difficult to estimate the number of sharks in the seemingly endless processions. As new individuals moved into sight, others slowly disappeared into the distance. (JAMES MCKIBBEN)

10) Two helpers and I spent a month during the summer of 1981 on Isla Pardito, a small island about ten miles from the seamount. We set up a small scientific camp next to the fishing camp. (Scott Michaels)

11) I recorded the behavior of the hammerheads within the schools using the first self-contained underwater video system. All of the earlier systems had a cable leading from an underwater camera to a recorder on a boat or on land. The housing containing the camera and recording deck was bulky, and provided ample exercise. (Howard Hall)

12) The whale shark, recognizable by its rows of iridescent light-blue spots and huge size, is the largest fish in the sea. It is a planktivore that feeds on the many shrimp, jellyfish, and larval fish born in the ocean currents. This massive shark has been observed to feed on small fishes by assuming a vertical position underneath a school and bobbing up and down while alternately opening and closing its mouth. (Flip Nicklin)

13) Southeast Farallon Island seen from Mirounga Bay. We watched for shark attacks in the waters around the island from the top of Lighthouse Hill, situated in the center of the island. (A. PETER KLIMLEY)

14) I am videotaping a white shark attack on a seal from the top of Lighthouse Hill. You needed to react quickly when you saw circling gulls or a pool of blood. (COLLECTION OF AUTHOR)

15) A white shark at the surface seizing a seal in its mouth. (PETER PYLE)

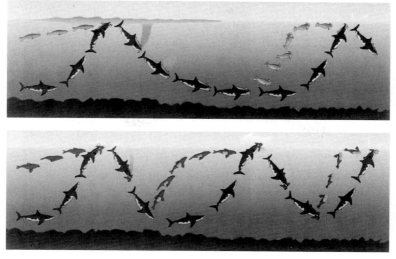

16) The sequence of events in a white shark attack on a seal differs from that in a sea lion attack. With seals (shown above), the shark seizes the seal and is able to carry it underwater until it loses most of its blood. It then removes a bite, letting the lifeless carcass float to the surface. Sea lions (shown below) often survive the initial bite and attempt to swim ashore, but are seized and bitten again, the second bite resulting in death. We often observed a splash at the surface during sea lion attacks, probably because the shark was seizing the sea lion with great force. (COURTESY OF THE SIGMA XI SOCIETY)

17) Cut Caudal (starting at top on the left) and Intact Caudal (on right) pass by each other twice while splashing water, in a complex ritual of competition for a dead seal. Intact Caudal is shown moving to the left and positioning itself between Cut Caudal and the seal, which is in the lower right corner of the front frame. The combatant that lifts more of its tail out of the water and splashes water the farthest is the winner of the ritual combat and is then able to feed on the rest of the seal. (COURTESY OF THE SIGMA XI SOCIETY)

18) This gray reef shark is exhibiting an aggressive display toward the diver as he approaches it. Note that the shark is swinging its tail far to the side, hunching its back, and lifting its snout upward. The diver had better retreat or the shark might turn around and attack him. (JAMES McKIBBEN)

19) A Mexican fisherman's daily catch of sharks stored in the bow of his panga. Few scalloped hammerhead sharks are now observed at the Espiritu Santo Seamount because intense fishing pressure has reduced the population in the Gulf of California. Sharks are more easily impacted by commercial fishing than other species of fish because sharks give birth to only a few live young and take a long time to reach maturity. (GIUSEPPE DI SCIARA NOTOBARTOLO)

island was the large, towering hill with the lighthouse on its rounded peak, surrounded by many other smaller hills. These made the island appear like asymmetric teeth on the cutting edge of a saw. As we approached the island from the east, we came to a sheltered bay, called Fisherman's Bay, with Lighthouse Hill towering above it on one side and, on the other, five or six massive rocks, each barely separated by water from the other. The farthest from the island looked like a loaf of bread and was called Sugerloaf. In the middle of the bay was a large buoy. "Ahh," I thought, "here is one of the two buoys that Dave told me about." That would be an ideal place to keep my panga during a storm because the bay was protected from the northwesterly winds and seas by the island and rocks on three sides. I noticed a narrow channel in the interior of the bay. Low, rounded waves periodically moved up the channel and broke on the beach, which was crowded with sea lions, standing upright on their two foreflippers and moving their heads back and forth in order to see us. The air was full of their doglike barking: *ark, ark, ark*. There was a flat rock ledge about halfway up the channel from which you could launch an inflatable and drive it to the buoy, to which the panga could be attached. You would want to get to the buoy quickly so that a large shark would not have the time to bite the inflatable and sink it, leaving you treading water at the surface. The shark would consider you easy prey as you swam ashore.

I asked the skipper to sail around the island so that we could identify landmarks on a topographic map of the island that Dave had given me (see map on p. 182). We sailed around Tower Point at the southern side of the bay and then passed Shubrick Point. Both of these points were ridges from Lighthouse Hill that led down to the sea. As we sailed around the eastern side of the island, we passed by the second buoy. Inshore of this buoy was a large channel, "Garbage Gulch" on the map, which led to a white sandy beach, littered with the large black shiny bodies of prostrate female elephant seals. The rounded heads of seals were also visible bobbing up and down on the surface of the water in front of the beach. The seals were continually making snorting sounds. You can make the same sound yourself by bringing your upper teeth down over your bottom lip and expelling air through the side of your mouth and forcing your lips to oscillate. It sounds something like a flatulent human being. Of course, the elephant seal was making the sound by forcing air through the fleshy proboscis at the end of its nose.

There on the southern side of the channel was a concrete platform on which was balanced a small Boston Whaler. The skipper told me that this was the skiff used to take persons as well as supplies to the island from the supply boats, which were tied up to the mooring. "Ahh," I thought, "a place for the wooden cradle that we built to hold my panga." Would there be enough room for the panga? Although slender, less than 7 feet wide, the panga was long, 23 feet from stern to bow. Next to the platform on the rocks was a smaller concrete platform that served as the base for the crane. The crane consisted of a long metal lifting arm composed of four metal girders, held together by smaller cross girders, and extending out over the channel. Cable passed from a large metal spool, situated near the platform, through a pulley suspended below the top of a tripod of girders and through a second pulley at the end of the lifting arm to a large pelican hook dangling over the water. The skipper told me that he would place his supplies in a wooden box aboard the Boston Whaler when it came out to the mooring. The box had cables attached to its corners that led to a metal eye, which could be slipped over the hook of the electrically powered crane, which would then lift the box up to the platform on the island. Soon a biologist appeared at the mouth of Garbage Gulch heading toward us in the Boston Whaler. In less than a minute he arrived at the sailboat. We lifted several boxes of supplies in the wooden box, and he drove off to take them to the island.

We entered a large crescent-shaped bay after passing between the shore and Saddle Rock. It surely looked like a huge saddle with a pinnacle at one end, a downward curving middle, and a broad peak at the other end, which was no farther than 100 yards from the island. We were now in Mirounga Bay, so named because it is the generic name of the elephant seal, *Mirounga angustirostis*. Many of these seals moved to and from the island in this bay. Barely visible above the waves and projecting 25 yards into the center of the bay was a small peninsula, 15 yards across at its widest, connected to shore by a narrow arch, 5 yards across, through which water constantly surged back and forth.

This was Low Arch, on which lay many juvenile seals that were the favorite meal of white sharks that frequented the waters surrounding the island. On the flat terrace in the center of the island behind the arch were two Victorian-style white houses with red roofs, separated from each other by no more than 100 yards. Each had a front door with a staircase that was covered by a small secondary roof. The doors looked directly out on Low Arch. The house toward West End served as the PRBO field station; the house toward East Landing was unoccupied. There standing in front of the field station was a biologist or volunteer, who was looking our way. I wondered at that moment whether more attacks were observed off the arch simply because it was directly in front of the house. The observers surely must spend more time looking in the waters in front of the field station.

At the western side of the bay and on the second island, West End, was another mountainous ridge that sloped down to the water. Seen from the bay, the outline of the ridge resembled the angular forehead of an American Indian, and hence was called Indian Head. There were no buildings on this uninhabited island. As we rounded this point and sailed along the southwestern coast of West End, the waves became larger and larger. We were getting closer to the northwestern coasts of the two islands, which were usually exposed to the roughest seas because the winds generally came out of the northwest. Here is the third bay, Maintop, a large crescent of shoreline stretching from the westernmost point of West End to the northern end of Sugarloaf. The skipper and I decided not to enter Maintop Bay. We watched the huge waves become steeper as they traveled closer to shore and formed steep crests that then crashed downward with white water, then splashing upward, all this with a deafening roar. This was a place to avoid. We turned the sailboat about and headed for San Francisco. Both of us were absolutely exhausted when we pulled into the slip at Pier 39 at dusk.

I started to search in earnest for white sharks at the Farallon Islands two years later, in September 1985, while on a three-day cruise to the island aboard the *Susan K*, the research vessel of the Bodega Marine Laboratory. Prior to that, during my second visit to Northern California in the fall of 1984, I had stayed overnight at a campground in Bodega Bay, a small town at the base of a small bay 80 miles north of San Francisco. We were there to search for sharks at Bodega Rock, a small island just outside of the inlet that was inhabited by harbor seals and California sea lions. That night, while we were pitching our tents at a campsite, a stranger drove up in a pickup truck, stuck his head out of the window, and asked for me. When I walked up to the truck, he told me that he was the administrative director of the Bodega Marine Laboratory (termed BML for short), a research unit of the University of California at Davis. He had read an article in the *Los Angeles Times* a few days before about my prior hammerhead studies and my quest to learn about the behavior of white sharks in the waters off the Central California coast. He later confided to me that he was impressed by my refusal to divulge where and when I would conduct these studies in order to work in privacy despite obvious media pressure. Asking me to join him in the truck, he drove less than a mile before we came upon a driveway leading to a lovely cluster of buildings, nestled against a hill, where visiting scientists could stay. He continued around the bay and turned onto the next driveway. It led half a mile to another, larger group of buildings—laboratories—situated on a shrub-covered plateau overlooking a bay with a crescent sandy beach called Horseshoe Cove.

He then took me to the municipal dock and introduced me to Deke Nelson, the captain of the *Susan K*. This was a 41-foot motorboat that had been converted to a research vessel by the addition of an A-shaped metal crane. Cable was passed through this A-frame and was used to drag nets and trawls to collect specimens for the laboratory. They encouraged me to use their research vessel to study white sharks. The vessel had a galley for cooking, a dining lounge, sleeping berths, as well as plenty of space in the stern for the large amount of baiting supplies needed to attract sharks. This vessel was to become my first base of operations in my quest to find white sharks in Central California waters.

My main objective now was to attach a beacon to a white shark and track it. I planned to do this by either of two methods. If the shark could be attracted close enough to the side of the boat, I would attach the tag to the shark using a long pole spear. The beacon would be mounted near the tip of the spear with a stainless steel leader leading to a sharp and pointed metallic dart that fit in the tip of the spear. I would have to reach over the side of the boat holding the spear in my hand with the rubber band from the base of the spear hooked on my thumb. I would then release the spear so it would spring forward and insert the dart into the thick muscle of the shark's back.

If the shark did not come close to the boat, I would have to resort to another method. The beacon would have attached to it a small treble hook. The electronic tag would be concealed within the carcass of a seal (or some other prey item) that was allowed to float at the end of a line attached to the *Susan K.* When the shark ate the seal, the transmitter would pass down its throat and the hook would attach to the lining of its stomach. The discomfort to the shark from the hook should be minimal because sharp, fractured bones of seals had often been found in sharks captured by fishermen.

First, we would have to attract a shark to the side of the boat. For that reason, the *Susan K* was loaded with stuff that would attract a white shark when we headed to the South Farallon Islands. We carried ground fish and blood to attract sharks, and several carcasses of sheep, in which we could conceal transmitters. We had been unable to find any freshly dead seals on the local beaches, so we decided to use the bodies of sheep, which we obtained from the Petaluma Livestock Auction. Once a sheep's skin was removed, it looked very similar to the white carcass of a seal floating in the water. We also had purchased a dozen 50-pound bags of ground fish from the Point St. George Fisheries, which supplied the ground fish to local fishermen to use in their crab traps.

A large amount of blood would be needed, and we obtained this at a local slaughterhouse. Collecting blood in a slaughterhouse was a very emotionally disturbing experience for me. The cows were killed, hung on a hook attached to a rail on the ceiling, and pushed toward the butcher, who would then skin them. She slit the throat of a cow, and told me to collect the blood that flowed out by holding my bucket underneath. My hands trembled as the blood filled up my bucket. Since each cow was pushed by me, still kicking wildly in post-mortem spasms, I had to be very alert to avoid being struck by a hoof.

In each 5-gallon bucket, I had poured 5 cups of vinegar to prevent the blood from coagulating. Slaughtering is a terrible but necessary consequence of our being carnivorous. Most of us are unaware of the process by which meat gets to the table. I tried to control my distress by reminding myself that the domestic cow was the second-most-successful species on earth, thanks largely to Homo sapiens, because the species was raised for its food value. Cattle are permitted to graze undisturbed in pastures until they reach adulthood, when they are dispatched quickly.

The trouble getting ground fish and blood led me to wonder whether I should substitute artificially synthesized chemicals for natural chum. In the late 1970s, a team of researchers had compared the attractive nature of various chemicals. The chemicals were placed in tanks containing sharks with electrodes leading from their head to an electronic instrument that displayed the electrical activity of the brain. Brain activity was highest in the presence of two common chemicals, glycine and betaine. So I obtained small containers of these two chemicals and brought them along with us to test their effectiveness in attracting sharks.

We left for Southeast Farallon Island from Bodega Bay at 6:30 A.M. on September 15 and arrived four hours later. This was my second cruise aboard the *Susan K.* A week before, we had gone to Bird Rock, a small island at the mouth of Tomales Bay, a shallow estuary a mile wide and 20 miles long along the San Andreas fault. Bird Rock was no more than 10 miles south of Bodega Bay, and had the dubious distinction of being that place where the most white shark attacks on humans had occurred along the West Coast. There had been seven attacks by white sharks on divers searching for abalone, a limpetlike shell containing delicious meat that attached itself to the rocks. Abalone feed on kelp, a marine algae common in these waters. We lowered perforated buckets and burlap bags of blood and ground fish into the water, letting the contents seep out continuously. We also mixed the two in a plastic garbage pail and used a restaurant ladle to toss some of the concoction into the water every five minutes. We chummed night and day for periods of twenty-three, twenty-eight, and five hours at different locations around the island without seeing a single shark come up to the side of the boat, to which we had attached a sheep carcass with a beacon stuffed inside it. The lack of success attracting sharks led me to question whether the imposing presence of the ship scared off sharks. Another factor to consider was that the ship's generator produced vibrations at sixty oscillations per second, which would surely be audible to the sharks and might frighten them.

For this reason, our first action upon arriving at Southeast Farallon Island was to set up two moorings with sheep carcasses attached containing concealed ultrasonic beacons. We placed one in the center of Mirounga Bay near Low Arch, which was now covered by the shiny brown bodies of small elephant seals basking in the sun. We could hear the seals' bleating calls above the thunderous crashing of the white waves. We dropped the other mooring close to Saddle Rock at the eastern edge of the bay. If a shark would not approach the *Susan K*, surely it would approach one of these sheep that resembled a seal carcass. Seal carcasses were common in the waters around the island and a likely source of food for the local sharks.

The problem with the mooring technique was to know when the shark took the ultrasonic transmitter. Deke, the captain of the *Susan K*, came up with a solution for this problem. He suggested that I place a radio transmitter on a shorter line leading to the sheep carcass being used as bait and then attach this line to the main line with a breakaway string so that the sheep (and transmitter) remained underwater. We would be alerted when the shark took the sheep because the radio beacon would pop to the surface, where its signal could transmit and alert us that the shark had eaten the sheep. Radio waves, unlike ultrasonic waves, do not transmit through salt water, only through air. We lowered a heavy anchor to the bottom and released the line with a buoy that floated at the surface. A shorter line leading to the sheep had been attached to the main line by a thin breakaway cord. The radio beacon was attached to the short line just above the sheep. The two bait stations were in place by a little after noon on the first day of our three-day expedition to the island.

We then left Mirounga Bay and anchored the *Susan K* behind Saddle Rock. We clamped the radio receiver's antenna to the roof of the boat and connected it to the receiver, which we had placed in the galley. We then set the controls of the radio receiver to alternatively scan the two frequencies of the radio transmitters, waiting in earnest to hear a high-pitched *blip-blip-blip* that would indicate that the shark had taken the bait. Then we ladled a mixture of blood and ground fish into the water around the sheep carcass, which had the third beacon inside it and was tethered to the boat. If a shark hit the sheep, we would see it.

It was about an hour and a half later, 2:13 P.M., when we heard a loud *blip-blip-blip* emitting from the speaker from the radio receiver. One of the radio beacons had risen to the surface, where it could now be detected. Everyone rushed over and looked at the panel of the receiver to see which mooring had been disturbed. Displayed in bright white letters on the panel was CHAN. 2, indicating the receiver was detecting the beacon on the mooring near Saddle Rock. We then looked in the direction of the rock, and just on the other side of the rock were close to fifty seagulls, shrieking loudly as they flew around in a large circle. We all knew that a shark had taken the bait.

Deke pulled anchor and quickly drove us around the rock and into Mirounga Bay. This took only seven minutes: it was now 2:20 P.M. There floating on the surface was the loop of line that earlier had been tied to the body of the sheep, yet the shark was nowhere to be seen. I now attempted to pick up the signal of the ultrasonic beacon, which was probably in the stomach of a shark swimming below us. I grabbed the earphones from the ultrasonic receiver and put them on my head. I then lifted the long metal staff with the hydrophone at one end and its direction-indicating handle on the other end over the side of the boat. I lowered the hydrophone into the water and slowly rotated it back and forth over a 360° arc, listening for the *bink-bink-bink* of the beacon. At first, I could barely hear the signal, but then it steadily became louder. After five minutes, the signal was not only very strong, but its intensity did not change as I rotated the hydrophone in a com-plete circle: the shark had to be directly underneath us. Then Deke yelled, "Shark up, on the surface!" I quickly lifted my head and looked in the direction Deke pointed. A huge shark, rapidly sweeping its mas-sive tail back and forth, rushed forward and opened its mouth wide, seizing and swallowing the 2-gallon plastic jug full of blood attached to the mooring. The shark then turned around and swam back toward the mooring, this time seizing and swallowing the burlap bag full of ground fish. Then the shark submerged, disappearing as quickly as it had appeared. I now listened again with the hydrophone, and to my frustration, did not hear the signal of the beacon. This was odd. The shark couldn't have moved a mile away during this time, beyond the range of the transmitter. Could the signal of the beacon be weakened when passing through the body of the shark, or (more likely) could the sound energy be absorbed by the fat of a seal eaten earlier? Only several years later did I arrive at a more plausible explanation—the shark had let go of the sheep carcass, which had floated to the surface, where the beacon couldn't transmit its signal into the water while drifting away from the island. However, I hadn't the faintest suspicion that sheep were not acceptable food for a white shark—after all, film documentaries portrayed them mindlessly biting almost anything, animate or inanimate, thrown in the water.

We circled the bay twice, stopping periodically so I could put the hydrophone in the water and listen for the transmitter's signal. We were then forced to leave the bay to anchor behind Saddle Rock because the winds had started to blow, creating huge waves that were now breaking in the bay. We were going to have to wait until the next day to search again for the tagged shark. On the following day, we circled the bay twice while listening for the shark without hearing its transmitter. We then anchored the boat where the mooring had been, chumming with blood and ground fish for eight hours, hoping the same shark might return. It didn't return, and rough seas forced us again to anchor behind Saddle Rock during the night. We returned to Bodega Bay on the morning of the following day, September 18, 1985.

We returned to the South Farallon Islands in the *Susan K* on September 24, less than a week after our first trip. We anchored close to Saddle Rock in Mirounga Bay and spent eighty hours continuously chumming in hope of attracting the same shark. Before we left, we circled the island slowly, stopping in each bay and listening for the *bink-bink-bink* of the beacon on the shark. There was no evidence that the shark was still in the vicinity of the Farallon Islands.

We made one more three-day cruise during the autumn of 1985 aboard the *Susan K*. We took the boat to Elephant Rock after a diver had been attacked the week before by a shark while collecting abalone near the rock. We used the same procedure as before, baiting from the boat and setting out bait stations with attached sheep carcasses. We chummed for twenty hours before proceeding to Point Reyes, a massive rock peninsula with beaches inhabited by seals situated halfway between Bodega Bay and San Francisco, where we had no success despite chumming for a similar length of time.

We had now made four three-day cruises to prime white shark habitat without attracting a single white shark to the proximity of the *Susan K*. We had chummed with massive amounts of blood and ground fish as others had done in South Africa and Australia. I asked myself, "What are we doing wrong?" Once again, I considered the obvious possibility that the sharks were avoiding us because they were afraid of the research vessel. Perhaps it was its size, or the generators producing vibrational noise, well within the shark's hearing range. Furthermore, it was not easy to take the *Susan K* to these remote sites, bait for several days, and then, if a shark were to be tagged, to immediately mount a long-term tracking effort.

I decided to search for the sharks from now on in my small, 23-foot outboard skiff, which would be unlikely to frighten the sharks. When I confided my plans to Deke, he shrugged and, with his quick sense of humor, reminded me of what the white shark did to the trawler in *Jaws*. Actually, I had little fear that the white sharks would try to sink my Mexican panga. It was as long as the largest white shark yet captured, and even one that size would likely consider us a competitor and not prey. The shark would probably perform some kind of noncontact aggressive display in order to frighten us and avoid being injured. It also made sense to be based at the South Farallon Islands. The likelihood of finding a shark there was the highest. Had we not attracted our sole white shark there? An assistant and I could stay at the field station on the island, searching for sharks aboard the skiff each day. If we succeeded in tagging a shark, the *Susan K* could be summoned and the shark could be tracked with the larger vessel. Finally, an island would be safer to work from than the coast of the mainland because I could seek protection from winds coming from any direction by driving the boat around to that coastline of the island sheltered from the wind.

On October 18, 1985, Deke took David Spinelli and myself out to the Farallon Islands aboard the *Susan K*. David was not only a smart young man, having recently graduated from the University of California at Davis, but was strong and physically fit, thanks to his having been a lineman on the football team. This was just the kind of person that I needed to work with in this rough-and-tumble study. The *Susan K* was loaded to capacity with scientific gear and bait for attracting sharks. Let me give you an inventory of our first cargo of shark attractants. The *Susan K*'s freezer had been filled with one 5-gallon bucket of blood, six 50-pound slabs of ground fish, and two sheep halves. Two 5-foot-long coolers were on either side of the boat: each contained a bucket of blood, two slabs of ground fish, and a burlap bag of ground mackerel. Between the coolers on the stern were seven 55-gallon plastic trash cans, each with a bucket of blood and three slabs of fish. In the center of the boat was our prized possession, a 350-pound frozen sea lion, which was wrapped in a tarpaulin and strapped to a stretcher. The veterinarian at the California Marine Mammal Center had donated this sea lion to our study after it had died of a disease earlier that month. We had stored it in the walk-in freezer of a restaurant in Bodega Bay.

When we arrived at Southeast Farallon Island, out to our boat sped the Farallon biologist Peter Pyle in the small Boston Whaler. Peter was of medium height, dark-haired, and had binoculars on a strap around his neck. He used the binoculars to identify birds in the waters around the island. He would become a collaborator for years afterward in my studies of white sharks at the Farallon Islands. He gave us the wooden box in which we would place our supplies. He then headed back to the island and was lifted aboard the small boat onto the island, where he would operate the crane. David and I loaded our gear into the wooden box and then drove over to the narrow channel over which the crane extended, put the eye in the hook, and let Peter lift the gear in the box onto the island. We made many trips to get all of our supplies onshore.

Peter was worried that the crane could not lift the panga, so we decided to tie it off at the buoy at Fishermen's Bay, on the eastern side of the island. He reminded me that we could drive out to the boat using a small inflatable that could be launched from the shore there. After I tied the panga to the buoy, we drove back to East Landing in his little 10-foot skiff. I was lifted ashore in a Billy Pugh; he was lifted ashore in the skiff. This is an interesting object, consisting of a canvas bottom attached to a metal hoop with four or five lengths of chain rising to an eye that is clipped onto the hook at the end of the cable from the crane. It is similar in function to a bosun's chair. It was an exhilarating experience to step onto the Pugh, hold on to the chains, be lifted 25 feet up along a steep rocky cliff, and brought down on the concrete platform at the edge of the cliff. That night at dinner, Peter told David and me that earlier that week there had been an attack on a seal by a white shark similar in size to the one we had tagged, not more than 100 yards off Saddle Rock.

On the following day, Peter showed David and me how to launch the inflatable from some flat rocks in the surge channel at the southern side of Fisherman's Bay. We then drove out to the skiff, started the engine, and drove around the eastern coast of the island to East Landing. Peter was waiting for us, standing next to the controls of the crane. He then proceeded to lower supplies to us in the box. We ended up with one of the white coolers filled with a bucket of blood and a couple of slabs of fish, two small moorings consisting of an anchor, line, and a buoy, as well as half a sheep and our prized possession, the dead adult sea lion. By midday, we had put in place two moorings, one 25 yards to the west of Saddle Rock with the half sheep attached and a second a similar distance to the east of the rock with the sea lion. We then anchored the boat next to the mooring with the sea lion and began to bait with blood and ground fish.

It was 1:30 P.M. when we started, and it was only two hours later that David said he saw something big swim under the boat. He thought that it might be a seal, but could a seal be 10 feet long? He couldn't see the whole body of the animal. I walked up to the bow of the boat and looked down at the sea lion attached to the mooring. The water was clear for Northern California, and I could barely make out the mottled gray colors of the rocky bottom. Then directly below me, into my view came the cross section of the body of a huge shark swimming directly upward toward me. I watched it for the next five seconds as it swam upward at a frantic pace, its mouth not open to reveal its teeth, until it was just below the side of the bow, on which I was standing. It looked at the boat, and possibly at me, before turning toward the sea lion carcass floating on the surface in front of the bow. It swam along the surface toward the carcass, looked at it, and then swam off. I was surprised that the shark didn't bite the sea lion. Why didn't it take the bait? Was it not hungry now because it had fed earlier during the week? Or did it prefer freshly killed prey.

On this second day at the Farallon Islands, I had the unique opportunity to observe the shark's predatory tactics from the perspective of its prey without being subject to the risk of an attack. I thought to myself how well the shark's approach is suited to surprise its prey. The shark's dark back made it difficult to see while it swam close to the dark, rocky bottom. It was only visible when it rushed upward toward the surface to seize its prey. It swam to the surface so quickly! I would not have detected the shark had I not been looking directly at it during that brief five seconds. Finally, the shark dashed toward its prey using a head-on approach, its 3-foot wide cross section being harder to see than its whole body, six times longer than its cross section. Several years later, while making a film about white sharks off Dangerous Reef, I observed two sharks perform this same dash to the surface over and over again as they repeatedly approached the boat from which we were baiting.

I immediately took out the telemetry receiver, put the earphones on my head, placed the hydrophone in the water, and turned on the switch to listen to see whether this was the shark that we had tagged almost a month ago. I was momentarily disappointed because there was no *bink-bink-bink*. However, I was excited by what I had learned about this species on my first day in the panga out at the Farallon Islands. What also impressed me was that the shark had arrived after only two hours, yet so little blood and ground fish had been dispensed that day compared to the amount we had used each day when we were operating from the *Susan K*. Was it possible that the boat did frighten the sharks and baiting was really not that important to attracting them? Could it be that only the sight of a seal or sea lion at the surface triggered a white shark dash to the surface to capture it? Chum, or the scent of prey, was certainly necessary to attract sharks to a pinniped colony from a great distance. However, once a shark was present at the colony, it might simply swim along the bottom, looking upward in order to detect a seal at the surface breathing between dives. We recovered our two moorings with their baits and drove over to North Landing, where we tied the skiff to the buoy, boarded the inflatable raft, and drove it to shore.

We were full of optimism on the next day, when we anchored the skiff and began to chum near Saddle Rock after setting out our mooring with an attached body of a sheep on the western side of Mirounga Bay. However, the seas were already choppy when we arrived, and the chop turned to huge rolling swells with crests extending across the whole bay. The swells were quickly growing in size, and breaking farther and farther from shore and closer to us. We had to stop baiting by noon after only an hour and a half. We then drove across the bay, taking care not to be caught on the crest of a breaking wave, quickly pulled up our mooring, and drove around the island to the flat waters, protected from the wind and waves, taking shelter in Fisherman's Cove. By 4 P.M. the winds were howling and seas crashing everywhere: a "northwester" was upon us and we were forced to return to the island. The violent storm lasted until midnight. Every two hours, one of us walked out to see whether our skiff was still afloat, tied up to the buoy in Fisherman's Bay. Huge waves rose up and crashed on Sugarloaf, the massive rock in the way of waves coming from the northwest. Early in the morning, the winds stopped as suddenly as they started: the storm had passed the Farallon Islands on its way eastward past San Francisco to the San Joaquin Valley and finally to the Sierra Nevadas.

The seas from the storm didn't settle down until early afternoon on the following day, the twenty-first of October. Needless to say, Dave and I were eager to get out to Mirounga Bay to find the white shark. David held a rope leading to the bow, keeping the inflatable pointed toward the incoming waves while I quickly jumped aboard the raft and, looking in the opposite direction, began pulling the starting cord of the outboard engine. He patiently waited for the engine to start so that he could join me and we could rush out of the surge channel between oncoming waves. Suddenly, a mountainous "sneaker" wave (much larger than the rest of the waves) loomed before us, swinging the inflatable around parallel to the crest, which grew higher and steeper, and then crashed downward, capsizing the inflatable. The engine and I were thrown high in the air, and we fell into the rough water, now rushing out to sea. My waders were full of water and pulling me away from shore. I used all of the strength I had to swim to shore. Soaked, I appeared in front of Peter Pyle, apologetic for what we had done to their inflatable. I blurted out that the inflatable had capsized, not sunk, but the outboard engine had come loose and had fallen to the bottom of the cove. Peter was sympathetic—he was well aware of the perils of launching from North Landing.

We now needed to retrieve the outboard engine quickly or our white shark studies might be over before they had barely started. I ran upstairs to my room, donned my wet suit, and grabbed my mask and snorkel. We were going to get that engine back! Peter and two volunteer biologists joined us as we rushed across the island to North Landing. I then jumped into the water between the crashing waves and starting looking for the engine while holding a rope tied into a hangman's noose. Wave after wave crashed down, pushing me forward and backward, and the white foam of the breaking waves made it difficult to see underwater. David Spinelli held the other end of the rope with all his might. I couldn't see a yard in front of me, and there might be white sharks here! Eventually, after thirty to forty-five minutes of searching, I caught a glimpse of the engine and was then able to slip the noose around the propeller shaft. Everybody on land took hold of the line and collectively pulled the engine ashore. We then took the engine to the foremen of a team of members of the California Conservation Corps who were temporarily staying in the second house on the island. He took part of the engine apart, cleaned the parts, and to our jubilant amazement got it running again.

Although we were successful in launching the inflatable on the next day, we were unable to drive the skiff to Saddle Rock because large waves were rolling in at Shubrick Point. It was not until the following day, the twenty-third, that we were able to anchor again at Saddle Rock and chum for sharks. I was standing on the seat in the middle of the boat with the pole spear holding the beacon in my hands, waiting for a shark to arrive so I could tag it. Attached to the stern by a thin breakaway cord was the thick line leading to my battery of attractants. They consisted of a burlap bag of ground fish, a plastic jug of blood, a small container punctured with small holes filled with a solution of betaine and glycine, and the carcass of the sheep. The purpose of the cord was to prevent damage to the stern if a shark were to drag the bait away from the boat. I momentarily shifted my gaze from the bait to David as I talked with him. At this instant, the boat suddenly lurched backward, the seat was drawn out from under my feet, and, having no support, I fell down on the floor of the boat. When I got up, I saw the line leading to the sheep floating now 30 feet away. A white shark had dashed upward to the surface, grabbed the sheep (with the transmitter inside), and swum 20 feet or so with it within its mouth, and had then spit it out.

This seemed odd behavior coming from what everyone believed was an indiscriminate feeder. I wondered out loud, "Why didn't the shark eat the sheep?" Did the boat frighten it? Was it because the sheep was attached to a line? The shark had dragged the line far enough away from the boat to break the cord attached to the transom. David let out the anchor line and I rowed closer to the shark, which was now swimming at the surface with its dorsal and caudal (tail) fins breaking the water. We couldn't get closer than 5 yards from the shark. Then the shark submerged, and swam under the boat. It was impossible, however, to reach down and attach the transmitter to the shark because it was 8 feet underneath the bottom of the boat. My inability to tag a shark was becoming very frustrating: white sharks seemed to avoid getting close to the boat.

David estimated this shark to be 14 to 15 feet long when it swam by the boat. He counted the number of meter marks on the waterline of the boat between the tips of the tail and dorsal fins. This distance was known to be a little less than half the length of the shark. The boat served as a tape measure, with thick black electrical tape placed every meter and thin tape every quarter of a meter along its length.

On the following day, David was injured when we loaded supplies into the boat. The boat suddenly dropped and moved forward with a passing swell. The large cooler suspended from the crane hit him in the chest, driving him backward. The seat in the middle of the boat pressed against the back of his legs, and he fell backward with his back landing on the floor of the skiff. The pain in his back was initially agonizing, and Peter and I discussed having the Coast Guard pick him up in a helicopter. However, he talked with his father, a physician, and they decided that he had suffered a muscle bruise and should recover in a matter of days. He persevered for the rest of the week, but the pain began to get worse. I was really worried about his health, and asked Peter how we could get him off the island as quickly as possible. Peter contacted the Coast Guard, which sent the buoy tender *Blackhaw* out to the island on the following day, to clean off the buoy at East Landing. At the same time, they picked up David and me. The *Blackhaw* took us back to San Francisco, where David immediately got medical treatment. In actuality, his injury was a muscle strain. This was verified after several X rays were taken of his back. Fortunately, the injury was not more serious—although it was very painful.

Less than a week later, on November 1, I returned to the island aboard a motorboat of the Farallon Patrol with a new assistant, Jim Wetzel, an avid diver from Bodega Bay with a keen interest in shark behavior. On our second time out after our return, we encountered a small white shark, 12 feet long, in Mirounga Bay. It swam under the boat three times, just deep enough to exceed the range of my spear. I bent over the side of the boat, extending my arm with the spear as far as possible under the water to get close enough to tag the shark. No luck! Again, the shark passed close to the sheep but did not swallow it. What was wrong with the sheep?

Upon arriving on the island, I asked Harry Carter, another Farallon biologist, who was taking Pete's place for the week, if we could keep the panga in the cradle on East Landing. (Harry was an expert ornithologist who studied the biology of the many seabirds on the island.) I was concerned that the shark might bite and sink the inflatable on our way out or back from the boat. We would be cast into the water and forced to swim ashore. The shark could easily attack us at this time, and without a mask and snorkel, we would be unable to see the shark. Because I was concerned about this danger, I stowed a pair of masks and snorkels in the inflatable for Jim and myself.

That night, we returned to the buoy in Fisherman's Bay, where we found Carter sitting in his skiff with the inflatable partly deflated and draped over his body and the engine underwater. A shark had attacked the inflatable between 1:15 P.M., when one of the volunteers on her rounds of the island making a bird census had noticed the boat floating at the buoy, and 3:45 P.M., when Harry spotted the inflatable sinking. Harry pointed at the pontoon on the right side of the rear of the inflatable. There on the pontoon were two huge crescents of 1½-inch slits made by the teeth from the upper jaw of a white shark. "Wow," I thought, "the shark must have been large!" The base of the crescent was 18 inches wide. If the shark were to open its jaws this wide, it could swallow a small human whole. The shark had bitten the inflatable exactly where I had been sitting earlier that day. "This was really a close miss," I thought.

Not only was there a crescent of punctures made by the upper jaw on the side of the pontoon, but also a crescent of punctures underneath made by the lower jaw. The position of the two crescents indicated that the shark had dashed to the surface at a steep angle. Two sets of upper and lower crescents overlapped this, suggesting to me that the shark had bitten its intended prey twice in quick succession, moving its head slightly to the side when biting down a second time.

We tried to salvage the engine that evening. We washed it with fresh water, immersed the parts in oil, dried the connectors in the ignition system with a hair dryer, and flushed salt water from the cylinders. However, we couldn't start it again despite our hard work. Later that year, the Park Service graciously provided additional funds for me to reimburse PRBO for the loss of their engine and inflatable raft. The part of the raft with the bites taken out of it still hangs on one of the walls of the field station on Southeast Farallon Island.

The next day, I measured the distance across the base of the largest crescent of tooth punctures on the pontoon and found it to be 45.8 centimeters (18 inches). It is possible to estimate the length of a shark based on the size of part of its body, such as the width of its jaws—there is a constant relationship between the dimensions of certain parts of the body and its length as a shark grows to adulthood. To estimate the shark's size in this way, you first need to collect measurements of that part—upper jaw width in this case—and the length from sharks that were captured and measured. The data for these sharks are usually in the field notes in fish collections. Then you can plot jaw width (the distance across the base of the upper jaw), versus total length (the distance from the tip of the snout to the tip of the tail), on a graph and fit a regression equation to the distribution of these points. You then enter a measurement of the jaw width into this formula and calculate an average estimated length of the shark of an unknown length. There will be a percentage of variability in the lengths of sharks with that jaw width. I had actually measured not jaw width but bite width—that part of the shark's wide-open jaw that left tooth impressions on the pontoon and, for this reason, my estimated size of the shark is slightly less than its true length. The average white shark with a jaw width of 45.8 centimeters would be 17 feet 1 inches long; individuals with this jaw width had been measured at lengths ranging from 13 feet 8 inches to 22 feet long. This shark could have been bigger than the largest white shark ever reliably measured by an ichthyologist off Cuba, which was 21 feet long. Since then, two substantially larger white sharks have been reported by fishermen, a 23-foot white shark off the island of Malta in the Mediterranean and a slightly larger shark caught off Kangaroo Island off South Australia.

Needless to say, whether 17 or 22 feet long, the shark that bit the inflatable was a large one. It was not by chance that the attacking shark was large: only a large predator would venture to attack the inflatable, which was close to 10 feet long. The inflatable had been a huge fishing lure, jerking back and forth like an animal swimming at the surface, as waves repeatedly drew it away and it recoiled toward the buoy to which it was attached.

We had our next chance to tag a white shark three days later. The shark circled the boat once at a distance, then came closer to within tagging range. However, just as I lowered the end of the spear with the beacon into the water and was about to release the spear, its other end hit Jim and could not be moved to the side. The shark was gone as abruptly as it had arrived; it did not feed on the sheep suspended from the transom of the boat. For some reason, the sharks were not eating the carcasses of sheep. This conclusion was reinforced when I pulled up one of the moorings and discovered that the burlap bag with fish had been removed by a shark, but not the sheep. Two days later, I retrieved another carcass that had small parts nibbled from it, but the majority of the carcass had been left intact. Obviously, the shark had put its mouth around the carcass, bit down lightly, and then let go. It didn't like something about the sheep. In this case, the boat was not nearby so it was unlikely that the shark released the sheep because the boat frightened it. I had by now become very frustrated by my inability to tag a white shark.

It was on November 14 that I finally succeeded. It was hazardous to lift our equipment and bait into the boat that day. Southeast winds were creating large waves that rolled directly into Garbage Gulch, where we loaded our supplies. We lowered the large cooler, heavy with bait, while our skiff moved forward and backward with each coming swell. You always had to keep an eye on the cooler because it could easily knock you off your feet. We drove the skiff to our favorite spot, anchoring the boat next to Saddle Rock in Mirounga Bay. We then started chumming. It was 10 A.M. Three hours later, a small shark approached the boat slowly, swam over to the sheep floating behind the boat, looked at it without taking a bite, and sped off in the direction of the oil slick leading from the bag of fish and jugs of blood and chemical attractants. When its tail was even with the transom of the boat, the shark's snout was even with the first small tape mark on the base of the boat following the four large tape marks: the shark was 4.25 meters, or 14 feet, long.

The shark returned within an hour and swam next to the boat with its back high out of the water and even with the top of the side of the boat. I placed my left hand on the warm body of the shark to find a soft, fleshy area in which to insert the dart with the attached beacon. I looked around to ask Jim for the spear and noticed that he was lying on the bottom of the boat with both of his hands placed over his head. This was scary stuff—one couldn't help being nervous as one of these sharks passed by the boat like a locomotive engine.

Everything that we had read in books and seen in the movies emphasized the dangerous nature of these huge predators. He quickly got up, grabbed the spear, and gave it to me. I then lifted the spear over the edge of the boat with my other hand, pointed the spear's tip holding the sharp metallic dart at the soft area, and let go of the spear so that the rubber sling contracted, thrusting the tip into the back of the shark. Then I howled at the top of my lungs, "We've tagged a shark!" Later, Peter said that my voice could be heard on the island, where he and volunteers stood holding their binoculars and watching the action before them. When the dart on the beacon was inserted into its muscle, the shark did not change its speed, indicating it felt little pain. Jim and I stood looking at this shark as it moved away, marveling at its size and power, the way you might stare in fascination at a locomotive passing you only a few feet away.

As soon as this shark disappeared below the surface, a second, larger shark surfaced near the sheep. We estimated that this one was 5 meters, or 16 feet 5 inches, long. This shark was easily recognizable by a crescent of white gashes inflicted by the propeller of a motorboat on its side near its dorsal fin. Jim and I could make out two large white organs trailing behind its bottom fin, and we agreed that these were claspers, indicating the shark was a male. The shark headed directly toward the sheep, grabbed it in its mouth, swam with it for a few seconds, and then spit out the carcass. The shark then submerged. Again, I was puzzled by the picky nature of the shark. I wondered whether the shark didn't like the line, which had attached it to the boat. We rowed the skiff over to the sheep and removed the line, which had broken away from the boat. Would the shark now eat the sheep?

Suddenly, up rose a *third* shark, a massive giant whose length we estimated to be 5.75, meters, or 18 feet 10 inches, almost as long as the boat. This shark swam up to the sheep, seized it in its mouth, and then spit it out and submerged. After a minute, this massive shark burst out of the water, propelling two-thirds of its body completely out of the water at a 45° angle to the sea surface like a ballistic missile leaving a submarine. The shark remained in the air for what seemed an eternity to us before crashing down on the surface and splashing water 20 to 30 feet in either direction. Much of this water landed in the bow of the boat. Jim and I stood there with jaws agape and eyes wide open, awestruck by the massive size of this creature and its impressive performance. It then crossed my mind that this might be an aggressive display intended to frighten us away from the bait.

By now the seas were rough, and we had to get back to East Landing. Jim tied the sheep with the transmitter to our anchor line and attached a buoy: perhaps one of the sharks would come back at night and eat the bait. We still hadn't given up using sheep.

On the following day, we circled the entire island and listened at two dozen sites for the signal of the tagged shark. The shark wasn't in Mirounga Bay, off Indian Head, in Maintop Bay, or Fisherman's Bay. The next day, a commercial fisherman took us in his trawler to the Middle Farallon Island, a round rock that protruded out of the water several miles north of the two South Farallon Islands, and the North Farallon Islands, a larger rock farther north. We couldn't find the tagged shark at any of these sites.

Three days later, the *Susan K* came to pick us up and take us back to the Bodega Marine Laboratory. My wife, Pat, was aboard the *Susan K*, and it was good to see her after being away for over two months. Before leaving, I took Peter around the island, instructing him in how to listen for the tagged shark at six locations. He continued to search for the shark after my departure, but never succeeded in finding it during the two more weeks that he spent on the island.

During this trip, a small white shark had swum up to our boat. This shark had an interesting color pattern: it was mottled with patches of light and dark gray. I thought to myself that the common name of the "white" shark was a misnomer. Although the fish would present a striking white belly as it was hauled onto the deck of a fishing boat, the name "black shark" would be more appropriate from an ecological perspective. An adult shark swimming close to the bottom would not easily be seen by a seal, because the black mottled back would blend in with the dark, rocky seafloor below.

I had tried to tag this shark with a streamer tag. Unfortunately, the shark kept at a distance from the boat. Some of the sharks seemed to exhibit the same behavioral tactic. They would circle the boat and attached baiting materials to learn more about the source of the chemical attractant. Then they would descend to the bottom, swim along it until they were below the prey, and then dash upward to seize an object at the surface. Pat and a young woman, volunteering as a biologist on the island, were present with us in the boat. It impressed me that everyone aboard the panga was not frightened by this shark, but excited about the prospects of tagging it and learning about its behavior. Fear changed to curiosity once people came in contact with the species. I was more determined than ever to learn more about the behavior of white sharks in order to provide the scientific community and the public with a more complete understanding of this awesome predator.

WHITE SHARK PREDATION
AT THE FARALLON ISLANDS

It was October 24, 1988, and we were standing on top of Lighthouse Hill, the pyramid-shaped hill, 340 feet above sea level, on the north side of Southeast Farallon. This is the best spot on the island for seeing sharks feed on seals in the surrounding waters. "There's an attack! Look there, over there," shouted Peter Pyle as he pointed at a large swirl of water mixed with blood close to Sugarloaf, the massive rock shaped like a loaf of bread at the northern edge of Fisherman's Bay. Peter looked down at his watch and then shouted, "The time is now 8:04 A.M." We were setting up equipment that would be used every fall for the next four years to document feeding by white sharks on seals and sea lions.

Lighthouse Hill is an ideal vantage point. To start with, on the peak there is an oval concrete walkway, 6 feet wide, that circles the conical lighthouse and has large aprons on its eastern and western sides, on which one can mount observational equipment. The equipment can be stored safely in the lighthouse at night, when it's not used to document shark attacks. It is easy to walk around this walkway, protected by metal railings on the outside, and look for attacks in the waters below. From here, one can look directly downward to the north on Fisherman's Bay, which is the home to a large colony of California sea lions. To the east is Shubrick Point, in the direction of the Golden Gate Bridge, which is often visible in the far distance. Groups of sea lions from Fisherman's Bay leave the island from this point at dusk on their nightly feeding excursions to forage on fish in the waters around the island, quickly escaping the perilous nearshore waters by swimming away with up-and-down surface dives like dolphins. Their behavior struck me as ideal for avoiding being surprised by a predatory shark lurking below—there were more eyes in a group to pick out a predator, and the members were moving targets with their porpoising actions. South of the hill is the expansive flat terrace on which the PRBO field station was situated. In front of the terrace is Mirounga Bay, crescent-shaped and the largest bay at the island. The bay is half a mile across. During the fall, many rounded and shiny bodies of juvenile elephant seals are present on the terrace, next to the small peninsula jutting out into the center of the bay. To the southwest is the smaller of the two South Farallon Islands, West End.

A large colony of seals is present during the fall on another flat area at the western edge of Mirounga Bay immediately below Indian Head. To the west is another large bay spanning both islands—Maintop Bay, which usually receives the brunt of huge rolling waves driven against the islands by the winds that constantly come out of the northwest. There is a large colony of elephant seals during fall on a large sandy expanse, Shell Beach, situated at the southern end of Maintop Bay and the northern end of the rock ridge on West End. Sea lions, on the other hand, cluster in great numbers high up among the rocks along the entire coastline.

From our high perch on Lighthouse Hill, we could see the entire rocky shoreline surrounding Southeast Farallon to the east and the coast of West End to the west. The weather was typical of that time of year, cold and foggy. Peter had spent many years on the island and was perfectly dressed for the weather. He wore a heavy cotton shirt, a pullover sweatshirt, and a baseball cap to keep the sun off his head. I had learned during my two visits to the island to dress for the weather, and wore an insulating fleece vest under a ski jacket, covering my head with a woolen cap. It could be cold on the hill.

Peter and I ran to the western apron of the concrete walkway circling the lighthouse, where we had the best view of Fisherman's Bay. Peter took hold of a spotting scope, a telescope mounted on a tripod and used for viewing distant objects, and rotated it toward the swirl of water. I grasped the handle of a video camera with a huge foot-long telescopic lens and aimed it toward the swirl of water, at the same time looking in vain for the edge of Sugarloaf in the viewfinder. I looked up to see the rock, and, finding the lens pointing to the right, moved the camera slightly to the left. "Ahh," I thought to myself, again looking through the viewfinder, "there is the edge of the rock—the camera needs to be moved a little to the right." My hand pushed the lens slightly, and there in the center of the viewfinder was a circular patch of smooth crimson water. I rotated a ring on the camera next to the telescopic lens and zoomed in on the site, making the patch of bloodstained water fill most of the viewfinder.

A shark had seized a seal swimming underwater seconds before, and blood was flowing out of the seal's body and coloring the surrounding water. Internal body fluids were also released and had covered the surface with a thin film of body oil, decreasing the viscosity of the surface and reducing wave action in the patch of water. The bloodstained oil slick was now moving toward the center of the bay. "Ah," I thought, "the shark is swimming below the surface with the seal in its jaws." As if reading my thoughts, Pete suddenly remarked as he backed away from the spotting scope, "They usually carry the carcass down, take their bite, and let the carcass float back to the surface." An endless time seemed to pass before the carcass floated up to the surface—thirty seconds—and oddly enough, it was not bleeding very much! This stuck me as strange, because seals have the largest volume of blood per body size in the animal kingdom. The blood carries oxygen to their tissues and enables young seals to stay underwater for periods in excess of twenty minutes.

Pete continued to narrate the attack in a loud voice, saying the seal was up and so was the shark, and giving the time. I remarked that one of the advantages of using our new camcorder was that time was continuously displayed in the lower corner of the video image. We wouldn't have to shout the time of each action in the future and could concentrate on carefully describing the behavior of the shark and its prey. As the seal floated to the surface and came into view, we could both see that it had been beheaded. That also seemed odd! I thought to myself: "How did the shark approach this seal close enough to bite off its head without being seen? Seeing the shark, the seal should surely have had the time to escape from the shark's onward rush. Perhaps, the seal had surfaced and lifted its head out of the water to breathe; during this brief moment of vulnerability, the shark might have been able to get under it, dash upward, and seize the unsuspecting seal as it put its head underwater again to dive. The seal must be 5½ to 6 feet long and weigh close to 200 pounds." Peter broke into my train of thought, saying, "It's a harbor seal—see its spotted fur."

I asked Jocelyn Aycrigg, a volunteer biologist helping that day with the shark watch, to take the camera and keep it pointed at the seal. I told her to keep the trigger on the handle depressed, and make sure the green light below the viewfinder was on: it indicated that the video camera was recording the attack. She looked through the viewfinder and remarked to Peter that she could now see a gull sitting on the carcass. Peter, an accomplished ornithologist, asked Jocelyn whether she had seen the rare glaucous gull that had flown past the carcass. She said that she hadn't seen the gull, but would keep looking for it.

It was important to record where the seal was first attacked. I now grasped a short telescope bristling with circular knobs on a second tripod, that stood next to the video system. This was a theodolite, an optical device that we would use to determine the direction and distance to the attack. Earlier that month, I had driven to Menlo Park, south of San Francisco, and picked up two of these sophisticated instruments from a scientist at the office of the United States Geological Survey. Earlier in the week, I had lugged both of these heavy instruments to the top of Lighthouse Hill. It was hard work carrying those heavy instruments up the hill. I had slowly walked up the long winding path, first leading to the left 100 feet, then cutting back to the right a similar distance, then continuing to the left as it led up the steep, rocky slope of the hill. We needed two theodolites, one situated on the eastern end and another on the western end of the walkway circling the lighthouse, to get bearings and ranges to attacks in the waters surrounding the coast of both islands.

I pushed the telescope on the theodolite around until a pointed cone on the end of the barrel of the telescope near my eye was centered in the notch between two cones at the other end of the scope. The blood-stained patch was now directly in front of the sight. I looked into the eyepiece (the ocular) and pulled the far end of the telescope downward with my hand until the crimson patch was visible. This process could be very confusing because the image seen through this vintage model of theodolite was upside down. I moved a mirror so it reflected sunlight into the scope. I now moved my eye away from the viewing ocular and prepared to look in the other ocular that would provide a direction to the origin of the patch of blood. First, I quickly rotated a circular knob so that a line crossing the face of the knob was horizontal, indicating that it was in the mode to provide a compass direction. I then slowly rotated a small knob in order to align the ticks on two scales, one on top of each, and read off the direction to the attack. I said aloud, "Zero-zero-three degrees, eighteen minutes and forty-one-point-two seconds," and then wrote this horizontal bearing on a form on a clipboard hanging on the tripod. I rotated the circular knob with the line so that it was vertical, looked through the small ocular, and read off the scale the vertical bearing to the attack. This was the angle between when the scope had been on a level plane and when it had been pulled downward to point at the patch of water. We would use this last number to calculate the distance to the attack using a bit of high school trigonometry.

The dead harbor seal floated at the surface for close to three minutes. Two and a half minutes later, Peter said, "OK—it's up again." The shark rose to the surface and slowly circled the carcass for twenty seconds before seizing it. The shark then carried the carcass in its mouth for forty seconds, propelled forward by the exaggerated movements of the tail, before taking a second bite, evident from the appearance of more blood in the water. Three minutes and forty seconds had now passed since the shark attacked the seal. Forty seconds later, Peter remarked while looking through the spotting scope, "One little piece of the carcass still on surface." The shark surfaced for the last time ten seconds later. It slowly swam up to the rest of the carcass, seized it, releasing more blood, and carried it for twenty seconds before swallowing the rest of the seal. I rotated the theodolite now to the final position of the shark and took horizontal and vertical bearings. It had taken the shark a little more than five minutes and three bites to consume the entire harbor seal.

The shark watch was long and tedious work. I got up before dawn and in darkness and hiked up the long path a third of a mile to the top of Lighthouse Hill. I then rushed to set up the video and two theodolites, one on either end of the oval concrete walkway, which was close to 30 feet across, before the sun rose. Here you had an unobstructed view of most of the coastline of Southeast Farallon and West End. All day long, you slowly walked around the concrete platform where, from a single position, you could see only a part of the coast and the surrounding waters. An average traverse of the platform took a minute. During this time, you looked for any evidence of an attack and checked out anything promising with a pair of binoculars or spotting scopes placed at the north and south ends of the platform next to the theodolites.

Shark feeding was often signaled by a combination of events. The shark's initial seizure of a pinniped—a seal or sea lion—was associated with a thunderous splash of the water or the sudden appearance of water that had been stained red by the seal or sea lion's blood. At all other times, you detected an attack shortly after its initiation. The dorsal and caudal fins of a shark might move along the surface of the water or there might be the motionless body of a pinniped on the surface. At other times, western gulls took off from territories over the two islands and converged over a particular patch of water. The gulls would circle while making high-pitched calls—*kee-kee-kee*—and individuals would fly down to the sea surface to feed upon small pieces of the prey's carcass.

You immediately dashed over to the theodolite and focused the scope on the place where the attack started. You then moved over to the camcorder, which was next to the theodolite, and began recording the behavior of the shark and its prey. During any break in the action, you read the horizontal and vertical bearings off the theodolite. You then took hold of the camcorder and recorded the behavior of the shark and its prey. When the action was over, you redirected the theodolite toward the place where the attack ended and took the two bearings. This all was usually done in less than ten minutes. It was indeed difficult to stay alert all day so that you would promptly detect an attack and fully document it before it was over. The monotony was broken every two hours when we carefully counted the elephant seals in each of twenty-four zones dividing the coastline of the island. We hoped to use this census to find out whether attacks were more frequent at those beaches where seals were more abundant, or if news of an attack spread around the island from zone to zone, causing seals to leave the water and seek the safety of the beaches. If you saw a seal or sea lion swimming away from the island, you took theodolite bearings and described its behavior. You walked around the lighthouse all day, stopping only after the sun set below the horizon and there was too little light to see a splash or the color of blood in the waters.

The shark watch evolved over time. After my initial stay at the Farallon Islands during the fall of 1985, Peter had asked me to make a data form with instructions on how to document the behavior of the sharks and their prey, a seal or sea lion, during a shark attack. During the following fall, he and the volunteer biologists on the island witnessed eleven predatory attacks while walking around the island and conducting their bird and pinniped surveys. They composed detailed descriptions of the events of an attack and provided a time for each event.

The volunteer observers also attempted to obtain the exact position of each attack using a handheld compass. This was not a very successful method for recording the position of an attack. To start with, one compass was not enough: they needed two compasses in order to record two bearings to the attack from widely separated locations on the island. The intersection of two bearings, or "lines of position," would indicate the site of the attack. Rarely did Peter and a volunteer have the time to walk to widely separated positions before the attack was completed. I realized that it would be far better to use a theodolite to determine the direction and distance to a distant object. A colleague of mine, Bernd Würsig, had set one of these optical devices on the towering cliffs along the Valdez Pensinsula in Argentina and used it to record the positions of dolphins and whales migrating along the coast.

However, it wasn't until the fall of 1987 that Scot Anderson, an energetic young man just out of college and a keen observer, further developed the idea of a shark watch. He came out to the island and spent the fall as a volunteer biologist. Peter wrote to me on October 3,

"Scot has been a great boon to our familiarity with Tiburon [the white shark]. He sat for about three–four days for most of the day on Falcon's Peak overlooking Mirounga and Maintop Bays looking for a predation event. He saw at least one attack a day and maybe twice he saw two. He works on fishing boats and has learned to watch the birds and the sea. One attack he cued into on the north side of Sugarloaf; he saw only two gulls headed in the same direction. He ran to where he could see where they were headed and there was a shark."

Peter went on in the note to say that it was clear that if we were actively to look for the attacks, we would see many more than had been observed serendipitously by people while walking around the island. He suggested that we assign a person full-time to this task next year. Scot filled out our forms in great detail, placing numbers over an axis with increments for each minute over a period of ten minutes and then for each number making a drawing of the behavior of the shark and seal. I called up the island by radio telephone and talked with Scot about the attacks. It seemed to me that we now needed a better way to document the attacks. I told him that I would send out a dictation recorder right away, on which he could record the time of the attack and then a narration of the events during the attack. Yet what we really needed was a camcorder with a high-power telescopic lens that could record both the voice of the observer but also the behavior of the shark and its prey. I knew we also had to procure a theodolite to obtain accurate positions for the attacks.

During the following summer, I acquired a video and a theodolite and brought them out to the island in October 1988. Scot's stay on the island that year overlapped mine, so I could show him how to use the camcorder and the theodolite. During the next four years, observers maintained a daytime watch throughout each fall for shark attacks in the waters surrounding the Farallon Islands. Over a six-year period from 1986 to 1991, they made 310 written descriptions and 131 videorecordings of predatory attacks. I collected the data forms and videos at the end of each fall season and devoted much of my time during each following year to analyzing them. Only through this collaborative effort, with field and laboratory components, were we able to unravel the secrets of white shark behavior at the Farallons.

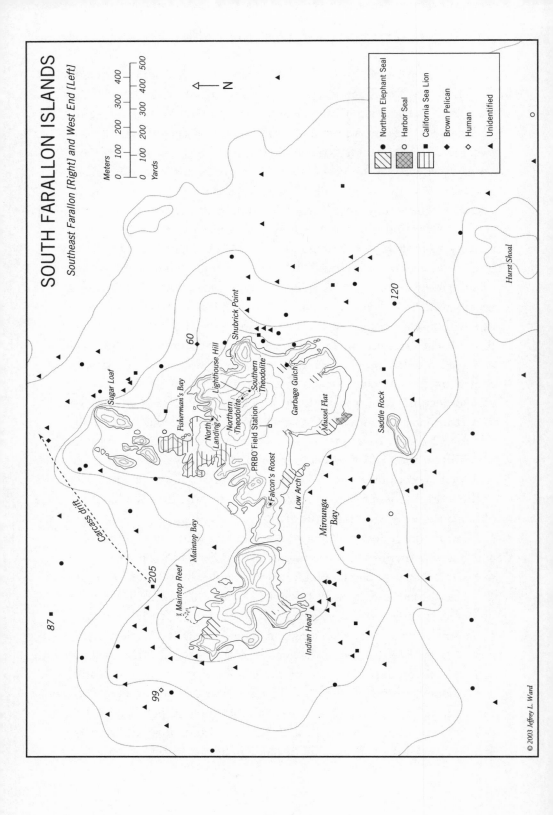

SOUTH FARALLON ISLANDS

Southeast Farallon [Right] and West End [Left]

Legend:
- ● Northern Elephant Seal
- ○ Harbor Seal
- ■ California Sea Lion
- ◆ Brown Pelican
- ◇ Human
- ▲ Unidentified

Meters
0 100 200 300 400

Yards
0 100 200 300 400 500

N

Hurst Shoal

Shubrick Point

120

60

Sugar Loaf

Fisherman's Bay

Lighthouse Hill

Southern
Theodolite

North
Landing

Northern
Theodolite

PRBO Field Station

Garbage Gulch

Mussel Flat

Saddle Rock

Falcon's Roost

Low Arch

Mirounga
Bay

Maintop Bay

Carcass drift

205

Maintop Reef

87

99

Indian Head

© 2003 Jeffrey L. Ward

I then processed this mass of data on a computerized workstation at my biotelemetry laboratory at the Bodega Marine Laboratory. Analysis of a record began by plotting the initial and final positions of the attacks. The first step was to find an accurate map of the Farallon Islands. A topographic map made by the Bureau of Land Management (BLM) had the best detail, showing not only the coastline at high tide but also the intertidal zone exposed during low tides. This map also had elevation contours so that I could accurately locate the position of the theodolite on Lighthouse Hill. I placed the map on a digitizer tablet and carefully traced the coastline of the island and elevation contours with a digitizing pen to produce an electronic image of the island. This became the initial layer of a file in a computerized drafting program. I would then enter information into additional layers, which could be superimposed on the map of the island. The BLM map had one serious shortcoming: there were no contours showing the depths surrounding the island. I found a nautical chart made by the National Oceanic and Atmospheric Administration that had far less detail to the coastline of the island but did show depth contours for 6, 10, and 20 fathoms. I reduced this map to the same scale as the other map, placed it on the digitizer, checked that landmarks along the coast so the two maps were aligned, and copied the depth contours around the island. I then plotted the positions of the predatory attack during each year on a different layer in this computerized file.

White shark attacks are sometimes considered random, hard-to-explain events, and some of the more hysterical ideas about sharks come from their portrayal as crazed, indiscriminate, and vicious

A map of the South Farallon Islands, with symbols indicating where white sharks have attacked different species. A number identifies each attack. The solid circles indicate attacks on seals, the solid squares attacks on sea lions, and solid triangles attacks on unidentified victims. There is cross-hatching at Low Arch, Indian Head, and Maintop Bay, the homes of large seal colonies. Notice that the symbols indicating predatory attacks are clustered at these points of arrival and departure from the elephant seal colony. In addition, there is a line of attacks leading away from Indian Head in a south-eastern direction, indicating the path to and from the island taken by elephant seals. The clear diamond labeled #99 off Maintop Reef on the northwestern corner of West End shows where Mark Tisserand was attacked by a white shark at 1:30 P.M. on the ninth of September 1989. (COURTESY OF THE AMERICAN SOCIETY OF ICHTHYOLOGISTS AND HERPETOLOGISTS) hunters. But at the South Farallon Islands, observations suggested that the attacks formed patterns in space and time.

First, although it is popularly believed that white sharks live mainly far away from shore, there is no denying that they frequently feed on seals and sea lions close to shore. It makes sense that they would maximize their chance of capturing their prey by staying near the places where these pinnipeds leave the water to spend more time in their colonies on land. Once they leave their coastal colonies, elephant seals migrate over great distances in the Pacific Ocean, where they are often separated from each other by great distances. It is when they come and go from their refuges on a few islands that they are concentrated that they are vulnerable to a predator such as the white shark. More than four-fifths of the predatory attacks seen at the South Farallon Islands were in a zone extending from 25 to 450 meters (27 to 492 yards) from shore. The volunteers at the island were able to distinguish dolphins, which were similar in size to the sharks, at much greater distances, so the zone's outer boundary does not seem to have anything to do with the visual abilities of the observers.

Within this zone, the predatory attacks were clustered at locations where elephant seals and sea lions entered and departed from the water. Furthermore, attack positions formed several linear patterns leading away from the island. Members of one of the colonies of northern elephant seals entered and left the water at Indian Head on West End Island, for example, and in a line extending southeastward from that location one could count eighteen recorded attacks. Of the six attacks where the prey was identified, five were attacks on northern elephant seals. A line of eight attacks extended offshore from Low Arch in the center of Mirounga Bay, and close to twenty attacks were clustered around Shell Beach on the southern edge of Maintop Bay. Both of these sites were entry points from large elephant seal colonies. There were close to twenty predatory attacks at Shubrick Point, where California sea lions left the coast of the island on their nightly feeding excursions.

The timing of the attacks also indicated some patterns. The rate of predation at the South Farallon Islands peaked well after dawn during the morning and slowly abated until dusk. We were unable to determine whether the sharks fed at night, since observations were impossible in the darkness. The retina in the white shark's eye has a high density of cone receptors (used for daytime vision) at its center, and a reduced number of rod segments (used for night vision), anatomical details that support the idea that the shark feeds during daytime. Not coincidentally, during the day seals and sea lions were active close to the shore of the islands.

It also turned out that predatory attacks took place at similar times on successive days. Scot suggested that the daily timing might be related to the fact that northern elephant seals are forced off the beach and into the water by high tides that occur at similar times on consecutive days. We found that there were more attacks at high tides, when seals were forced off the safe beaches that the sharks patrolled around the island.

It also appeared that seals and sea lions try to avoid attack when passing through the high-risk zone. Shark attacks often took place near the sea surface, and rarely do northern elephant seals surface within this attack zone. Sea lions usually pass quickly through the zone, porpoising like dolphins in tight groups. It is possible that this acrobatic behavior is an evasive tactic, and the group swimming is a kind of pinniped buddy system.

I noticed a difference between the way sharks attacked seals and sea lions, the two common inhabitants of the Farallon Islands. This was not unexpected, for their two favored prey species act very differently in the water. There were two species of seals that commonly inhabited the Farallon Islands, the massive northern elephant seal and the smaller harbor seal. Seals have well-developed hindflippers, which they use to propel themselves forward through the water. They swam to and from the island close to the bottom to avoid being surprised from below by sharks. The sharks fed most often on one- or two-year-old juvenile seals that were captured as they passed through the high-risk zone near the colony. It was very rare that a seal on the way back to the island tried to get a breath of air at the surface, but it was then that an individual was most vulnerable to attack.

Sea lions use their large foreflippers to propel themselves through the water in a fashion comparable to the way birds fly through the air.

The sea lions often moved to and from the island in large groups, porpoising through the water like dolphins. There were two species at the islands, the California sea lion and the large Steller sea lion. California sea lions were attacked by sharks much less frequently than the northern elephant seals; the large Stellar sea lions were rarely attacked. This may have been due to their large size, and their ability to bite or to rake the head of the sharks with the sharp claws on their foreflippers. Sharks filmed underwater at the island often have wounds or scars on the snout near the mouth. There may be a substantial risk of injury when a shark seizes a large California sea lion or Steller sea lion and attempts to consume it. Even a large northern elephant seal may bite a shark while held in the shark's mouth. Consistent with this risk is the high frequency of adult male northern elephant seals that survive attacks and swim ashore at Año Nuevo Island with shark-bite wounds.

The next step was to describe the behavioral tactics used by the sharks to capture seals and sea lions. We had videotaped the behavior of the sharks and seals during 131 of the attacks. I now had to describe each type of behavior shown by the sharks and seals. I had done this before for the scalloped hammerhead shark: making an ethogram (catalogue of behavioral patterns) of the behaviors performed by individuals in the schools around seamounts in the Gulf of California.

Next, I needed to develop a form of written notation to show others the sequence of behaviors by the predator and prey during the attack. My need was similar to that of the musician, who prepares a musical score composed of circles, indicating tones, that are placed on a staff consisting of a series of parallel lines, with each line, or space in between, designating a different pitch. Another musician could look at the score and play the song. I needed a form of notation to convey the sequence of events in an attack to others in a scientific paper without showing the reader the video, just as the musician provided a score without playing the song to a listener.

I decided to represent each behavior of a shark or its prey in one of two ways, as an "event" or a "state." There are some actions that are discrete and short in duration that are best described as events. Think of the shark's sudden seizure of a seal in its jaws from below. I called this action *Vertical Bite* in my white shark ethogram and defined it as an event:

Vertical Bite (event V): The shark swam upward from beneath the prey and grasped the prey in its jaws at the surface. The angle of the shark's body to the sea surface ranged from 60° to 90°.

There are other actions that are continuous or repetitive and longer in duration. These are most often locomotory movements such as the shark's swimming at the surface with wide tail beats. I classified this action as a state in the shark ethogram and referred to it as *Exaggerated Swimming*.

Exaggerated Swimming (state SWX): The shark swam by moving its tail back and forth over a wide arc, often displacing water with each beat. When swimming in this manner, the shark was believed to carry a pinniped (seal or sea lion) in its mouth. The exaggerated tail beats appeared necessary to propel the shark forward when carrying a bulky carcass.

I needed to express the sequence of the behavior on a scale of time. With an electronic drafting program, I made an axis of time consisting of small horizontal bars, alternating high and low, with each equivalent to ten seconds. Short vertical lines were added to the time axis at one-minute intervals. Above the horizontal axis, I recorded the behavioral acts of the shark; below, the actions of the prey. I would slowly advance the video until a behavioral event, such as *Vertical Bite*, was visible on the monitor, and would draw a thin vertical line in the space for behavioral events and identify it with a letter denoting the behavior. I drew a horizontal bar in the space for a state such as *Exaggerated Swimming* and extended it from the start of the state until its end. It was time-consuming, but important, to convert each video of a predatory attack into this type of diagram.

I was initially depressed while viewing the videos depicting the violent attack of sharks on the seals and sea lions. I may have related to the desperate situation of the prey because pinnipeds like humans are mammals. Yet after a while, I began to look at the videotapes more dispassionately. They could just as easily depict predation by a smaller species from another taxon, such as a spider. The behavior of the sharks was really no more morbid than a spider catching a fly in its web, injecting poison into the fly with a bite, and later feeding on the fly.

The sequence of actions in a white shark attack on a seal was often the same. For example, take predation P-120 on a northern elephant seal. This attack occurred at 3:08 P.M. on October 15, 1989, 350 meters southeast of Garbage Gulch. The location of the attack is shown on the map of the South Farallon Islands on pasge 182. Observers were alerted to the shark's initial strike by the appearance of a small bright red area on the sea surface so I placed a B for *Blood-stained Water* above the zero under the time scale on my diagram. The best-known event in a shark attack, the powerful first bite, was not usually witnessed by observers, suggesting the shark initially seizes the seal under the surface. I slightly rotated the large knob on the analysis projector, slowly increasing the speed of the video until the shark rose to the surface and began swimming with exaggerated tail beats. I thought to myself again that these sweeping tail beats would be necessary to propel the shark forward if it were carrying the heavy seal in its jaws. I looked at the counter: thirty seconds had elapsed since the appearance of the pool of blood. I used the computer mouse to place a cross-hatched bar over a tick on the time axis for thirty seconds and lengthened the bar until it was over the tick for forty seconds to indicate that the shark had performed *Exaggerated Swimming* for ten seconds. The bloodstain on the surface of the water had elongated in one direction. The seal rose to the surface after a minute and a half and floated there motionless while no longer bleeding. The seal was just outside a corridor of bloodstained water and 75 to 100 meters away from the site of the initial strike. I used the mouse to place a second clear bar, indicating the state *Floating Motionless*, in the space underneath the time axis for the behavioral states of the seal. The seal had a deflated appearance, having lost a massive amount of blood. These initial events, similar from attack to attack, are consistent with the theory that the white shark kills its prey by exsanguination, or blood deprivation. It appears that the shark holds a seal tightly in its jaws until it is no longer bleeding and then bites down, removing a chunk.

I advanced the video again while looking at the monitor. The carcass remained at the surface only ten seconds before the shark seized it again and submerged. Splashing and a small amount of blood appeared a minute later near the head of the shark, indicating that the carcass had been bitten again. I stopped the video, looked on the counter for the time that had elapsed since the beginning of the attack, and used the computer mouse to add the behaviors to my attack diagram. I added a thin line, the abbreviation SA indicating *Splashing at Anterior of Shark*, and B below indicating *Bloodstained Water.* I then advanced the video slowly again, while using the computer mouse to put four V symbols on the diagram indicating that the shark seized the carcass from below four more times. The elephant seal was consumed in eight minutes.

A typical attack on a California sea lion was videorecorded at 8:45 A.M. on November 13, 1988, 400 meters north of Maintop Reef. The observers on Lighthouse Hill were alerted to the attack by an explosive splash. The sea lion surfaced quickly and swam in a slow and jerky manner, often changing directions. It was traumatized and swimming with difficulty because the shark had removed a large bite from its flank. The sea lion submerged after ten seconds, then rose to the surface and again began to swim futilely toward shore. The white shark then abruptly propelled itself out of the water, seizing the sea lion in its mouth and coming down headfirst with the side of its tail hitting the water surface and creating an immense splash. Scot Anderson, recording this behavioral event from Lighthouse Hill, was impressed by its awesome nature. He remarked: "There is the prey swimming. It's still alive. Oooh! The shark came up and got it. Wow! Wow! He came right up from underneath and bit the hell out of it. [Pause] Umm. That was a little impressive!" The sea lion was never seen again.

Unlike seals, sea lions were usually capable of swimming after a shark had removed a bite from them. However, the shark soon returned to seize the sea lion in its jaws, carry it underwater until the bleeding stopped, and then take a second bite. The sea lion, now dead, floated to the surface, where it remained until the shark returned to finish it off.

Although observations of the shark's behavior were converted into a graphical representation, the sequence of behavior exhibited by the predator and prey was not the same in every attack. The next step was to arrive at a model, or generalized description of an attack, based on many more attacks on each of the species. There were fifty-five videos on which the prey could be identified to be a seal, and thirteen videos on which the prey was observed to be a sea lion. I recorded the relative frequency of each behavioral event and state in the first through sixth places in the sequence of acts in all of the attacks on seals and sea lions. The first six events were chosen because they were judged the most important during an attack. I made two graphs, one for interactions between sharks and seals and the other for the interactions between sharks and sea lions. Referring to the percentages in these graphs, I was able to make quantitative comparisons such as 77 percent of the seal attacks started with B, *Bloodstained Water*, versus 22 percent of the sea lion attacks. Or 54 percent of the sea lion attacks began with S, or *Explosive Splash*, relative to only 8 percent for the seal attacks.

This information permitted me to develop a conceptual model depicting the most common sequence of behavior during a predatory attack on a seal or sea lion (see figure 16 in photo section). An example, though arbitrarily chosen, represents the actual sequence of acts during an attack, yet the order of acts may differ from feeding bout to feeding bout. On the other hand, a model does not necessarily contain the sequence of events in any particular attack, but represents a sequence of acts that follow each other with the highest probability. These are two different perspectives on the same behavioral process.

I now had a good idea of the shark's normal predatory behavior after watching the many videos of sharks feeding on seals and sea lions. However, the sharks did behave oddly on some videos. Take for instance attack P-205 on September 10, 1991. The behavior of this shark first led me to wonder whether white sharks were picky feeders rather than the mindless feeding machines depicted in the movies. On this day, a high tide dislodged a dead sea lion from the beach, and the carcass floated out into the shark attack high-risk zone in Maintop Bay. The body of the sea lion was very buoyant because it was filled with the gases produced during decomposition.

Upon seeing the carcass, Peter remarked, that it "didn't look very appetizing." Normally, a white shark would vigorously consume the carcass of a freshly killed sea lion: seizing it in the mouth and carrying it awhile before removing the second or third bite. This decomposing sea lion was allowed to float near the island 132 minutes without a shark showing any interest in it. When a white shark did surface, it hesitated from feeding on the sea lion as if agreeing with Peter's assessment of its foul condition. The shark slowly circled the rotting carcass, inspecting it closely. Then the shark submerged briefly and surfaced underneath the carcass with its jaws held open, momentarily grasping the carcass in its jaws before releasing it without removing any flesh. The shark did this repeatedly—seven times over a period of seven minutes—before it finally approached from the side and slowly took a bite. The shark then swam away, leaving the carcass to eventually drift out of sight and away from the island. I noted on the attack diagram: "This shark had a unique feeding manner. Until the last pass, it moved only its head out of the water to mouth the prey, and never removed a chunk of prey. This shark is a discriminating feeder!"

A white shark also behaved in an unexpected manner toward a human being during attack P-99. This was the attack on Mark Tisserand, a commercial diver. The shark seized Mark in its jaws at 2 P.M. on September 9, 1989, after making a dive 250 meters west of Maintop Reef. He had surfaced and was swimming back to the boat carrying abalone that he had collected from the bottom during the dive. The shark and Mark were not recorded on video because our view was blocked by hills along the coastline. However, I was able to construct a diagram of the attack from a telephone conversation with him and the notes of Bob Lea, who had interviewed him for his file of shark attacks off the western coast of North America for the California Department of Fish and Game.

The attack started with the same behavior seen in a feeding bout on a seal. Mark told me that "the shark came up from underneath" so I put a V for *Vertical Bite* on the diagram for the initial behavior of the shark. Bob wrote, "Blood. A lot lost!" so I placed B for the initial act of the prey. Mark told me that "it carried him down for five to seven seconds," so a bar was placed on the graph for the state *Carrying* that was seven seconds in duration. *Carrying* was defined as: "The shark swam with an immobile or struggling pinniped [or human in this case] held in the jaws." Up to now, the shark's behavior toward the human resembled its usual behavior toward a seal or sea lion. Now came the critical difference between this attack and others on seals and sea lions. Bob told me that Mark had struck the shark with the butt of a bang stick, a behavioral event not witnessed in other attacks and so defined as *Strike with Object*. According to Mark, "the shark suddenly let go and swam off." I added the symbol R for *Release* to complete the diagram. This behavior, recorded during the first and only attack on a human during the period I was acquiring data, had been recorded only twice before, when sharks had released sea lions when feeding on them. In these instances, the shark did not get a firm hold on the sea lion's body and the sea lion was able to swim to shore despite its bleeding wound.

The fact that Mark hit the shark with the butt of a metal bang stick might have induced the shark to let go. (A shark's head has mechanoreceptors, or tiny pressure receptors, that make it sensitive to any contact with a hard object, and its electroreceptors are also sensitive to the galvanic fields emitted by metals underwater.) But the pattern of behavior that Mark reported—being bitten and released—is typical of white shark encounters with people off California. Some of these people did not have an object on them with which to strike the shark, yet the shark still spat them out without removing a bite.

Did the shark release Mark because he struck it with an object? Or could there be another reason for the shark to spit him out? I looked through our predatory attack diagrams for an example of an unresisting animal that was not consumed by a shark. I found it in attack P-60 on a pelican on September 6, 1998, 150 meters north of Shubrick Point. The pelican had just alighted on the water and was slowly paddling toward the island within the high-predation-risk zone when observers at the Lighthouse saw a thunderous splash. The shark seized the pelican briefly but then quickly released it. There on the attack diagram, the pelican was indicated to be *Moving Limbs*, a state of a dying animal, "floating on either its back or its belly and feebly moving its limbs up and down." The bird had been disabled and was unable to resist further attack. Next, the pelican spent two minutes *Floating Motionless in Bloodstained Water*. At this time, the shark rose to the surface and swam by the pelican but never ate the bird even though it was easy prey.

Finally, there was the odd behavior exhibited by sharks toward sea otters. One day during spring 1983, Jack Ames, a biologist with the California Department of Fish and Game, had led me into a room with a dead sea otter lying in the middle of a table and pointed to a crescent-shaped scar across its back. He then took tweezers and pulled out of the large wound a tiny white fragment and placed it under a microscope. I looked into the scope, and there along the edge of the shiny and polished surface of this hard object were serrations resembling those along the cutting edge of a steak knife. The sea otter had been attacked by a white shark! This was part of a shark's tooth that had broken off when the shark had seized the sea otter in its jaws. White sharks and other predatory sharks have as many as a dozen rows of teeth for seizing and biting off pieces of their prey. These teeth are shed when a shark feeds and either fall to the seafloor or, as in this case, remain embedded in the flesh of a potential prey victim seized by the shark. This was just one of many sea otters with scars containing fragments of white shark teeth that Jack had collected from the beaches around Monterey Bay. I recorded in my field notebook on March 17, 1988, that "possibly, the wounded otters are only those which the white sharks failed to capture. However, sea otters have yet to be found in the gut of a white shark." Even then, I was perplexed at why these apparently nutritious sea otters were not eaten by sharks.

An explanation of why the sharks might be picky feeders came to me after reading an article on an entirely different topic: hunting by prehistoric man in North America. I read with interest an article published December 1993 in *Science News* about a study by anthropologist John Speth of the University of Michigan. He was characterizing the diet of early hunters based on his excavation of a series of prehistoric deposits in southeastern New Mexico that contained the bones of bison butchered around A.D. 1450. The remains at the site, perplexed Speth because the hunters had a selective appetite. They left female bodies to rot at the butchering site, yet carried home as much of the male carcasses as they could manage. He wondered what was wrong with the female bison. The answer was the absence of fat. He wrote: "The wildlife literature said what was obvious in hindsight, that pregnant and nursing cows are often severely stressed in the spring because they are carrying a full-grown fetus or nursing a newborn while the forage has yet to start growing. So they live off their fat reserves and get fat-depleted."

The article went on to note that their body fat can decrease to only a few percent of their body weight during the cold and dry seasons when they start to starve. The fat can be less than what appears in even the leanest cuts of steak. A diet made up of almost pure protein would contain too few calories and would lead to protein poisoning, and this may have explained why the hunters rejected the extremely low-fat meat of the female bison.

Speth recounted the experience of a military officer who ran out of food in southwestern Wyoming during the winter and marched his men all the way back to Santa Fe, New Mexico, to find provisions. In his diary, the officer wrote, "We tried the meat of horse, colt, and mules, all of which were in a starved condition, and of course not very tender, juicy, and nutritious. We consumed the enormous amount of from five to six pounds of this meat per man daily, but continued to grow weak and thin, until, at the expiration of twelve days, we were able to perform but little labor, and were continually craving for fat meat."

After reading this, I wondered, "Could white sharks also have a craving for fat?" This surely would explain the anomalous behavior of the sharks exhibited toward some species. The birds, sea otters, and people that the sharks rejected were composed mainly of muscle, whereas the seals and sea lions that the sharks ate were composed mainly of fat. The energy content of fat, 8.0 kilocalories per gram, was close to twice the energy value of the protein in muscle, 3.9 to 4.5 kilocalories per gram. Could the sharks be testing these items for fat while carrying them in their jaws? Surely a shark's jaw would penetrate more easily into the soft external fatty layer of a seal than into the tough, muscular body of Mark, who was in peak physical condition.

Consistent with this theory was the manner in which a white shark fed on a dead whale floating at the mouth of San Francisco Bay. Deke Nelson, captain of the *Susan K*, had given me a film of the incident. Sweeping its tail back and forth, the shark propelled itself forward onto the carcass of the whale with its jaws open wide. The shark then lowered its upper jaw, penetrating the fatty outer blubber layer of the whale with its teeth. The shark then swung its tail and bent its rear body backward, the resistance of the water enabling the shark to drag its jaw down the body of the whale, stripping out a large scoop of fat. I had looked over the length of the whale and had noticed that only the outer fatty layer had been removed. Could the sharks prefer energy-rich marine mammals such as seals, sea lions, and whales, to other comparatively energy-deficient species? The juvenile elephant seals were particularly high in energy value; almost half of their body mass was composed of fat. These seals would be virtual power bars to sharks.

I thought to myself that there was certainly irony in the evolutionary history of the relationship between this prey and its predator. Seals had succeeded in colonizing the cold waters of the temperate and polar regions by evolving a thick layer of insulation. The white shark, unlike the bulk of shark species adapted to warm waters, had been able to follow the seals into these cold waters because it exploited the high energy contained in the fat from the insulation layers of seals.

The question remains, Why should sharks have a preference for fatty prey? The answer may be that surplus energy is needed to keep the shark's body warm. I remember placing my hand on the back of the large white shark that moved like a locomotive along the side of my boat. Its body was warm in contrast to the cold water off Central California. I later confirmed this observation by recording a temperature averaging 75°F with an ultrasonic transmitter inside the stomach of Top Notch, a large female, while she searched for seals in mid 50° water near Año Nuevo Island. As I noted earlier, in the body of the white shark are retes, or clusters of blood vessels, which warm the brain and musculature, increasing the transfer of oxygen to the tissues and quickening the reaction times of the nervous system in the cold waters inhabited by the sharks. Perhaps high-energy food also contributes to the maintenance of the shark's warm body temperature. Furthermore, the high-energy diet helps the shark grow fast. Adult white sharks increase their body size more than 5 percent a year. This growth rate is twice that of the porbeagle shark, a fish-eater in the same family that similarly inhabits cold temperate waters, and three times that of the mako, another fish-eater from the same family that lives in warm tropical waters.

I wondered whether in my laboratory notes there might be some quasi-experimental evidence for a feeding preference in white sharks? One way to test this theory was to alternatively present sharks with fatty and lean baits. A few years before, I had been perplexed by the unwillingness of white sharks to feed on sheep. This was the major obstacle to my tracking white sharks at the Farallon Islands during the mid 1980s. On November 4, 1985, I wrote, "Why didn't the shark feed on the sheep?" Two days later, I drew a picture of the rear end of a sheep showing only two small bites and tooth marks where the shark had seized and released the carcass twice. On November 14, I wrote, "Two sharks had taken the sheep in their mouths, swam away, and they released it. Perhaps its being attached to a line prevented them from feeding on it."

Contrasting with this lack of interest in sheep muscle was the sharks' interest in seal fat witnessed by myself, Scot Anderson, and Ken Goldman while a graduate student at San Francisco State University. In 1998 I was able to induce Top Notch, to swallow seal and whale fat in which transmitters were concealed at Año Nuevo Island. Ken and Scot eventually succeeded in tracking white sharks at the Farallon Islands. They towed a surfboard at the surface. When a shark surfaced on four different occasions to investigate the board, they tossed into the water a large chunk of fat with a hidden beacon. In each case, the shark swallowed the chunk of fat. Of even greater significance was the response of a shark to the rib cage of an elephant seal from which the fat had been removed. In correspondence with me, Scot Anderson had noted that "the shark bumped once, and left. After several minutes, I retrieved the line with the uneaten bait." The great white shark has through a fat-consumption preference, staked out a cold-water realm. I could not help thinking about how the people of the arctic regions thrived on a fat-rich diet. Life is a constant search for energy— the key to survival.

CHAPTER 11

TALKING WITH THEIR TAILS

It was 9:30 A.M. on Saturday, March 6, 1993, the day before my forty-sixth birthday. I stood behind a wooden podium next to a film screen in the lecture room at the Bodega Marine Laboratory. I was dressed in blue jeans and a shark tee shirt, as has always been my custom, but on this formal occasion I was also wearing a navy blue blazer. The room was dimly lit except for the bright white screen behind me that was illuminated by the slide projector.

Looking around the room, I recognized the familiar faces of scientists from all around the world who had come to Bodega Marine Laboratory to share what they knew about the white shark. David Ainley, the biologist with the Point Reyes Bird Observatory who had encouraged me to go to the Farallon Islands in 1983, and I had organized this four-day symposium on the biology of the white shark. We had sessions with different talks reporting on the species' evolutionary history, physiology, behavior, ecology, population biology, and interactions with man.

There sitting in the back was Otto Gadig, a biologist from the Universidade Federal da Paraíba, who had reported the rare appearance of white sharks off the coast of Brazil not far from Rio de Janeiro. Sitting not far from him was Malcolm Francis, a fisheries scientist from New Zealand, who had earlier described courtship among white sharks off the South Island. One shark had bitten another, the two eventually became motionless, one under the other, and then they turned over lying on their backs belly to belly. Next to him and closer to me was Senzo Uchida, head aquarist at the Okinawa Expo Aquarium. He had talked about oviphagy, or uterine cannibalism, in the white shark—the propensity for babies to eat each other while in the female's womb. In the sand tiger, a species related to the white shark and in the lamnoid order, only two siblings are produced per litter, one from each uterus, because each sibling devours all of its rivals, fertilized eggs and smaller embryos within the uterus. Sitting even closer to me was Barry Bruce, a member of the CSIRO, the fisheries agency in Australia. He had warned us of the decline in the population of white sharks near Dangerous Reef and the Neptune Islands in Spencer Gulf, a large bay in South Australia. Across the center aisle was Geremy Cliff from the Natal Sharks Board of South Africa, who had provided an estimate of the number of white sharks swimming in the waters off South Africa. In the first row was my good friend Len Compagno, imposing with his great bulk and powerful intellect, an American scientist who had described in minute detail the cartilaginous elements in the skeleton of the white shark. He was author of a compendium on the sharks of the world and was presently curator of fishes at a natural history museum in Cape Town, South Africa.

Don Nelson, friend and mentor since my hammerhead days, was moderator of this particular session, which focused on the behavior of the species. Standing at my side, he now introduced me: "Dr. Klimley will now describe two displays, the *Tail Slap* and the *Breach*, and intraspecific competition among white shark, *Carcharodon carcharias*. His coauthors in this paper are Peter Pyle and Scot Anderson." Peter and Scot sat in the middle of the room with the American contingent of researchers. I began slowly, reading my notes. Until now, I said, we scientists had been talking about the white shark's mouth and jaws. I paused a moment and said that my topic would be the shark's tail. In the previous talk, I reminded the audience, we had described predatory behavior of an individual shark. Now it was time to describe intraspecific competition among individuals, largely based on repeated observations of competitive interactions between two sharks that frequently came to Southeast Farallon Island. Intraspecific competition occurs when two or more individuals of the same species simultane-ously demand use of a limited resource, in this case an energy-packed seal. Access to a resource such as this is commonly established among animals through aggressive behavior that rarely takes the form of direct fighting. Competitors instead resort to displays. These are con-spicuous and exaggerated postures and movements of the body that show the animal's uneasiness at the presence of another and the ani-mal's capacity to inflict harm should that competitor remain. The sig-naler will gain an advantage if the recipient of the signal heeds its message and withdraws. Did you ever hold up clenched fists and bare your teeth to intimidate a bully in your class who wanted to take something of yours? That is a human display.

Now think of two white sharks attempting to feed on the same seal. It would be disadvantageous for one to discourage the other by biting it and inflicting a wound. This is because the situation is recip-rocal—the individual inflicting the bite this time may receive it next. Such an injury would reduce the ability of either of the two sharks to catch seals in the future. For this reason, a "display" of one's potential would be advantageous to the white shark. Avoidance of injury during intraspecific confrontations while mating or feeding accounts for the evolution of threat displays in the majority of animal species.

My first objective was to describe competition among white sharks feeding on seals floating at the sea surface after being killed by the first strike and in turn outline the postural elements of a threat display, the *Tail Slap*, and the social context in which it occurs. A slide was now projected onto the screen showing the tail of a white shark lifted high out of the water and ready to be brought down upon the surface of the sea. I explained to the audience that 98 *Tail Slaps* were observed during 26 of the 131 predatory attacks recorded on video from 1988 to 1991. In 23 of these 26 attacks, the prey was observed to be a northern elephant seal. The behavior consisted of a shark lift-ing the caudal fin out of the water, pausing as if to direct it in a partic-ular direction, and then rapidly lowering it while contacting the water with considerable force, often splashing a large amount of water in the direction of another white shark.

On the screen now was a slide composed of a dozen frames of

video showing the positions of the tail fin during a *Tail Slap*. I pointed at the frames on the screen, explaining that the motor pattern could be broken down into the following actions. While swimming at the surface, a shark first rotated on its side. Then it bent its body at a point approximately two-thirds of the distance from the snout to the tip of the caudal fin, lifting the tail upward and out of the water. Elevated to the maximum height, the tail formed an angle with the sea surface ranging from 30° to 90°. The size of the angle depended upon the degree of rotation of the shark's torso: a small angle resulted if the shark tilted only slightly, a large angle if the shark turned almost completely on its side. The time interval between when the shark initially lifted its caudal fin out of the water and when the fin was brought down to contact the sea surface was only 0.6 seconds.

I then briefly described three predatory attacks in which the participants performed *Tail Slaps*. The examples were chosen to illustrate the range in complexity of the behavior. The first consisted of a single shark slapping its tail twice in the direction of another shark. The second was a spectacular, rapid-fire exchange of *Tail Slaps* between two sharks swimming next to each other. The third was an elaborate ritual comprised of two white sharks passing each other repeatedly swimming in opposite directions and striking water toward each other as they passed.

A symbolic representation of attack P-140 was projected on the screen. Contained within a rectangle was an axis for the time elapsing during an attack, and above and below were lines and bars leading to letter abbreviations for behavioral events and states of the shark and prey. I raised the wooden pointer in my hand and pointed to the line leading to "SL." This was an abbreviation for the event *Oil Slick*, defined as the sea surface becoming smooth due to the dampening effect on small waves exerted by oils escaping from the body of the prey. I explained that the members of the shark watch at Lighthouse Hill were alerted to this predatory attack a few hundred meters northwest of the coast of the western island by the presence of an oil slick with a small bloodstain in the center. A white shark surfaced two minutes and twenty seconds later and swam slowly within the oil slick. The shark then began to swim very vigorously, sweeping its tail back and forth over a wider arc and powerfully propelling itself forward. If you looked at the shark's tail, I pointed out, it was completely intact with its upper caudal lobe visible. This shark, which I referred to as Intact Caudal, probably held the prey its mouth. It then submerged, but was observed swimming intermittently at the surface until twelve minutes into the attack. At that time, a second shark surfaced and performed two *Tail Slaps*. This second shark was easily identified because it was missing the upper lobe of its tail, and is appropriately named Cut Caudal. After the two *Tail Slaps*, Intact Caudal submerged and was never seen again. In contrast, Cut Caudal continued to swim at the surface, where it was later observed to bite the carcass. I now played a video of the attack while pointing out the two *Tail Slaps*. It was always necessary to show a video after presenting a symbolic representation of an attack—beyond the cerebral appreciation of the coded diagram of the attack, even the most serious scientists were galvanized by the visceral shock of seeing the awesome behavior of the shark.

A diagram showing predation P-143 was now projected on the screen. I quickly described this second example of communication among sharks. Five days later, the same pair of sharks had again fed in the same place along the coastline. Observers were alerted of attack P-143 by the appearance of blood at the sea surface off West End. An immature northern elephant seal floated to the surface after 1 minute 40 seconds with a bite removed from its midsection and remained lifeless within a pool of blood for 1 minute 20 seconds before Intact Caudal surfaced and seized the seal. Later, Cut Caudal appeared at the surface. At this time, Intact Caudal lifted its tail out of the water perpendicular to the sea surface and brought its tail downward twice, displacing large amounts of water. Then Cut Caudal lifted its tail upward and brought it downward against the sea surface, displacing even more water. This was immediately followed by a third *Tail Slap* by Intact Caudal and a second *Tail Slap* by Cut Caudal. The five *Tail Slaps* occurred in rapid succession, being completed in only 2.84 seconds. The two sharks were in such close proximity that it is likely that they struck each other during this behavioral exchange. Only Cut

Caudal was observed afterward, carrying the carcass of the seal in its mouth at the surface. Intact Caudal had been chased away. I then showed a video of the rapid succession of *Tail Slaps* with huge amounts of water splashing everywhere. There were oohs and ahs from several in the audience.

Next was a complex ritual in which both sharks exhibited the display. This third predatory attack involving the *Tail Slap* was observed a year later. At this time, Cut Caudal and a second shark (based upon size, likely Intact Caudal) repeatedly swam by each other at the surface. I raised the pointer and directed it to "SS," the abbreviation for the behavioral event *Side by Side*. As the sharks passed by each other during the second and third passes, each shark rolled over slightly, lifted its tail out of the water, and brought it down against the sea surface, splashing water in the direction of the other. After doing this, the two sharks immediately turned around and passed by each other a second time, again splashing water at the other. Then they did it a third time. During all three passes, Intact Caudal positioned itself between Cut Caudal and the seal carcass in order to prevent the latter from feeding on the seal. Later, during this third encounter between the two sharks, Intact Caudal surfaced and fed on the rest of the prey. This bout of *Tail Slaps* is illustrated in Figure 17 in photo section.

I then presented evidence indicating that this behavior was a social display. Multiple sharks were observed during attacks with *Tail Slaps*. Two or more sharks were observed during seventeen of twenty-six of the attacks with *Tail Slaps*. In four of the other nine, attacks were either preceded or followed by attacks with two sharks. The observers on Lighthouse Hill may have missed a second shark, although it was likely present during these other attacks. Multiple sharks visited the surface more often and lingered there longer than single sharks during attacks with *Tail Slaps*—an extra time investment necessary to discourage a competitor from feeding on its prey. I then concluded the talk by stating that the inference that the *Tail Slap* was an intraspecific threat was supported by the context in which the behavior was performed. The tail was at times brought down to strike the sea surface at such an angle that a large splash of water was propelled toward another shark. Furthermore, the displaying shark often positioned itself between the carcass of the prey floating at the surface and another shark, thus preventing it from feeding. The *Tail Slap* usually elicited one of two responses by the second shark. Either that shark withdrew from the area, permitting the displaying shark to feed on the prey, or it performed a *Tail Slap* in the direction of its neighbor. In this latter case, which of the two sharks eventually fed on the prey more often depended on the vigor and frequency of the *Tail Slaps* than the size disparity between the sharks.

Don Nelson now walked behind the podium, turned toward the audience, and asked whether there were any questions for me. Len Compagno, who was sitting in the first row, quickly stood up and said excitedly, "We have seen splashing, rather exaggerated splashing, when sharks feel resistance from bait held on the end of a rope. This is a similar thing, but the splash is not directed at us." Don then wondered out loud, "Where is the dead seal? Is it possible that the shark is simply trying to maneuver into a better place to bite the seal?" The next question indicated further skepticism toward the idea that the sharks were communicating. "Couldn't this behavior have simply been an artifact of the power burst at the surface, where the tail has no place to go but become airborne. I have seen this in a tank when they are feeding."

I thought a moment and then replied to the last question: "I know that this is a possibility. There are times when a shark might grab the prey, attempt to push it down, and the resistance of prey could force the shark's tail to come out of the water. However, I do not see that that could explain what is shown here on the screen." I backed up the slides one by one to the collection of video frames depicting two sharks lifting their tails and splashing water at each other as they passed by each other. I then directed the pointer at the carcass of the prey and said: "See here, this shark is staying between the carcass and its competitor to prevent it from feeding on the rest of the seal. In some cases, I might be wrong, but in others a communicative function seems likely to me. Your alternative might explain a single slap of the tail, but not the many *Tail Slaps* shown on this slide. That was the purpose of showing you bouts of the behavior with increasing complexity."

Not everyone in the audience was convinced that the *Tail Slap* was a display behavior. This sentiment became more apparent during an informal session of talks held that evening on the predatory strategies of the white shark. I gave a short critique of the "bite and spit" theory of prey capture. According to this theory, the shark seizes a seal or sea lion and then releases it in an intact yet wounded condition, waiting nearby until the wounded prey became immobile and could be eaten without risk of injury. This theory was based on records of sharks attacking humans and then leaving them uneaten, not on actual observations of sharks feeding on seals. I suggested that my theory of prey capture—exsanguination was a more common tactic to incapacitate the prey. There were no questions about the "bite and spit" or "exsanguination" hypotheses, but there were many questions about the *Tail Slap* display.

Greg Cailliet, a professor specializing in fish ecology at the Moss Landing Marine Laboratories, said that he for one remained unconvinced that the slapping of the shark's tail served the purpose of communication. My explanation of the motivation behind *Tail Slap* seemed to him unnecessarily complex. Perhaps his opinion was a product of his particular expertise, determining growth rates in fishes. These rates were determined precisely by counting the number of alternating rings on a shark's vertebral disk, which was stained with a dye and appeared light and dark under a dissecting scope when illuminated by ultraviolet light. He said emphatically, "You need to worry about what was in the shark's mouth, because the slap of the tail may result from the shark pushing the prey downward." I replied, "But the prey is often on the surface, and not in the shark's mouth when the shark performs the *Tail Slap*. The caudal fin is also lifted in the air and rotated in order to direct water toward the other. It would be hard to accomplish this with the prey in its mouth." Greg still was skeptical: "The shark could be doing that [lifting its tail], but for some other reason. You haven't eliminated other explanations!" Greg's attitude reflected the rule of parsimony—the simpler explanation is usually correct when evaluating two hypotheses, one simple and another much more complex.

The rapporteur, the person who was later to summarize and pro-vide an overview of the results presented in this session, was George Barlow, a professor who taught classes in animal behavior and fish biology at the University of California, Berkeley. He was an eminent animal behaviorist who had served as editor of the journal *Animal Behaviour* and had published several articles on communication among fishes. He now suddenly stood up, walked in front of the blackboard, picked up a piece of chalk, and began to give us an elo-quent and succinct lecture on fish communication while illustrating it on the board. He began by saying: "Two fishes in an aggressive encounter approach each other head-on. Like two cars playing chicken, they turn off at the last moment and end up parallel to each other but facing in opposite directions. You can see this between two dogs; you can also see this entire sequence repeated with two school-children on the playground. What they are doing is positioning themselves so that they can turn toward or away [from their adver-sary]. They are feeling a conflict between attacking and fleeing from their opponent. What commonly happens in fishes is that they tail-beat one another: each fish beats its tail to propel water toward its adversary. To prevent itself from moving forward, the fish holds its side fins outward and exerts pressure in the opposite direction. Now if a shark were to engage in tail-beating, it would have difficulty turning its pectoral fins into a brake position. However, if its tail end were out of the water, then it would derive little propulsion from it and would not be thrust forward. In addition, the tail could generate an acoustical signal by hitting the surface of the water—and we know that sharks hear well."

George continued: "Sharks have a difficult problem communicat-ing. Like most fishes, sharks can't move their fins, change their color, or produce sounds. Their means of communication are limited. One of their few options at the surface would be to perform this kind of behavior—a *Tail Slap*. So, in my view, if there are data indicating that the *Tail Slap* is directed toward another shark, Peter has prima facie evidence that this behavior acts in communication. It is incumbent on those who feel otherwise to come up with an alternative hypothesis that better explains this behavior, such as turning away to escape from the opponent. But you have to come up with a consistent hypothesis as an alternative, not just protest your disbelief." With that said, George walked away from the blackboard and sat down.

We had by now been debating the function of the *Tail Slap* for over an hour. Bob Lea, custodian of the shark-attack file for California, now rose and commented that the debate was very interesting and could easily last for an hour more, but why not continue it in a more casual setting, such as the lounge while drinking a glass of wine or beer. This ended a lively debate among some strongly opposed to the idea and others embracing the idea.

Walking out of the auditorium, I thought that it would take all of my savvy as an ethologist to change the minds of colleagues so res-olutely opposed to my interpretation of the *Tail Slap*. The task would even be harder because my perception of the white shark as an intelli-gent creature capable of communication contrasted greatly with the widely held perception that the shark was a dumb solitary species that fed in a simple, reflexive manner. I decided then to send the manu-script containing our results for review by animal behaviorists, who would be familiar with the concept of animal communication and likely sympathetic reviewers, but also to send it to these vociferous critics, who were trained in other disciplines and skeptical about the existence of animal communication. The challenge was to address their concerns as well as those of the behaviorists.

The debate over white shark tail slapping raged late into the night. At some point, some of my colleagues decided to create a parody of my scientific presentation. I inadvertently aided them in this endeavor. In my role as an editor of the conference proceedings, I sent to every-one a short version of my article on the *Tail Slap*.

The format of my colleagues' article was identical to mine, except it was entitled, "The Evolutionary Significance of the *Tail Flap* Behav-ior in the White Shark: Alternative Hypotheses." The article included the usual sections for a scientific article: an abstract, introduction, materials and methods, results, discussion, and acknowledgments. The purpose of the parody was explained in the abstract. They wrote that Klimley et al.'s consideration of the significance of *Tail Slaps* in white sharks prompted a reexamination of this behavioral event. They offered two alternative hypotheses to the interpretation of the behavior as ritualized combat. One hypothesis stated that the *Tail Flaps* (not *Tail Slaps*) between upper caudal lobes were congratulatory responses by a pair of white sharks to successful predatory events involving pinniped prey. The other hypothesis suggested that *Tail Flaps* were part of ritualized displacement associated with a pair of male white sharks courting a female. Displacement behaviors seem out of context. They are commonly exhibited by animals in situations characterized by conflict: a human blinks the eyelid or taps the fingers when nervous. My colleagues boasted that their hypotheses were pretty darn good.

The origin of the article was mentioned in the materials-and-methods section. They wrote that one of the alternative hypotheses was developed following the evening section, entitled "Predatory Strategies of White Sharks," and began in earnest during a celebratory event, initiated by the Refreshment Committee (RC) of the conference. My colleagues were now using the same method of abbreviation for behaviors that I had used in describing the predatory and communication behavior of white sharks. They stated that their study had involved approximately three hours of intense and often heated deliberations, accompanied by more than sufficient Liquid Sensory Stimuli (LSS), and so on. Many ideas were exchanged during this discussion, resulting in as yet untested hypotheses, which they were to report within their scientific paper.

They then illustrated their version of a predatory attack that lead to the successful sharks performing a high-five with their tails, termed a *Tail Flap*. In this diagram, however, the behavioral events and states were not devoid of human connotations such as mine. They chose letter abbreviations such as SM for *Serious Maiming* and MP for *Major Pain*; for states, they chose LBS for *Laid-Back Swimming*, SLH for *Swimming Like Hell*, and BIG for *Basking in Glory*. And of course, there was TF for *Tail Flap*. This behavior was apparently known as HF for *High Five* in North America or OC for *Oi-cinco* in the Brazilian literature. (I now knew that my Brazilian colleague, Otto, took part in the parody.) For the first figure of the paper, they made a sketch of two sharks holding their tails out of the water and hitting one against the other in a *Tail Flap*, or a *High Five*, shark-style.

A diagram was included in the paper showing their version of the behavior of the shark and its prey throughout the attack. This was the third figure in their paper. They referred to the great white as PDG or "well, at least a *Pretty Darn Good*" shark. They stated in all seriousness that Figure 3 depicted a *Fairly Normal Sequence* or FNS involving predation by PDG white sharks on their pinniped prey. Typically, two males teamed up to thin out an already overloaded population by utilizing "bite and spit" or exsanguination attack patterns (they wouldn't take sides because they both sound PDG) to *Seriously Maim* or *Kill* a poor, warm brown-eyed HS or *Huggable Seal*. The next two behavioral events were designated on the diagram by lines leading to the abbreviations for *Seriously Maim* and for *Dead Pinniped*. They went on to say that the sharks first observed a potential victim, indicated by a line leading to ELS on the diagram denoting an *Early Lunch Sighted* and were observed themselves by the pinniped, indicated by a line leading to FL for *Fear and Loathing*. All three parties then changed from *Laid-Back Swimming* into *Swimming Like Hell* modes. Following a successful attack resulting in *Serious Maiming*, *Major Pain*, and *Dead Pinniped*, both sharks are seen to exchange *Tail Flaps* as part of the *Basking in Glory*. Invariably, the dominant male, characterized by the presence HW for *Hard Wangers* gets in more Flaps than the immature male, which has SW or *Soft Wangers*. (This last sentence was a reference to Wes Pratt's talk at the symposium in which he discriminated between immature and mature male white sharks on the basis of the hardness of their two claspers, the intromittent organ inserted into a female to make her pregnant. Male white sharks become sexually mature at a length of 13 feet 2 inches, when they first develop rigid claspers.)

Later that year, David Ainley, my partner in editing the book, sent our article to four reviewers, including the colleague most outspoken in his criticism of the work during the plenary session on white shark predatory strategies. In the collective opinion of the reviewers, we needed to address three concerns in order to convince them that the tail was used in communication. First, we needed to demonstrate that the tail did not simply break the water because the shark was redistributing its weight to push the massive body of the seal downward.

This was an attempt to make us conform to the rule of parsimony—the simpler of two hypotheses explaining a phenomenon should be favored unless that hypothesis is rejected on the basis of sound evidence. Second, it was essential to provide evidence that the magnitude of the motions of the shark during a *Tail Slap* were greater than during *Exaggerated Swimming*, when the shark carried the prey in its mouth. Finally, the reviewers wanted us to demonstrate that the display intensity of the winner, which consumed the rest of the seal, was greater than that of the loser, who was unable to feed anymore on the seal. After reading the four outside reviews and David's appraisal of them, I was thunderstruck by the Herculean effort that must be expended to address these important concerns. After sitting in my office a couple of hours considering alternative plans of action, I finally outlined in my laboratory notebook the types of measurements needed to answer these challenging questions.

My first objective was to show that the shark did not simply lift its tail out of the water to redistribute weight in order to push the highly buoyant seal in its mouth downward. I needed to know, first, whether the shark seized the prey before the *Tail Slap*, and second, whether the shark submerged afterward. I sat down in front of my computer, opened up the first year-long file of the predatory attacks (from 1989), and began looking through each diagram of a predatory attack with a *Tail Slap*.

The first step in analyzing each diagram was to look immediately in front of the line leading to the *Tail Slap* for evidence of the shark seizing the prey. The behavioral combinations indicated that the sharks did not submerge after a *Tail Slap* as though trying to push their prey downward. I counted thirty-three times when a shark did not seize a pinniped, performed a *Tail Slap*, and also remained on the surface. In twenty-eight of these, two sharks were present at the surface during the *Tail Slap*; in the other five examples, a second shark was observed later during the attack. This says something about how the sharks communicated. If the shark preferred to strike the water only when its opponent was at the surface, the opponent might be expected either to see the tail raised in the air or feel the water splashing against its side. In contrast, there were only three examples of the shark seizing the prey, performing the *Tail Slap*, and then submerging. In these few cases, the tail may have come out of the water when the shark was redistributing its weight to push the prey downward. However, these examples were rare. There were also many examples of a shark seizing the prey, performing the *Tail Slap*, and staying at the surface. This led me to wonder whether holding the prey in the mouth might be an integral part of the display, enabling the shark to raise its tail higher out of the water and splash water farther. There was no doubt of the social nature of the *Tail Slap*. It was recorded only during attacks in which two sharks were identified, never during a similar number of randomly chosen attacks in which only one shark participated.

The next issue to be addressed was whether the slap of the tail was actually an exaggerated motion. It would now be necessary to measure various aspects of the display. I chose to measure four components: the amount of body lifted out of the water, the duration that the tail was lifted out of the water, the distance water was splashed when the shark's tail contacted the sea surface, and the time that the water remained in the air. All of these might indicate to the opponent the physical prowess of the displaying shark. The magnitude of these behavioral elements needed to be compared to the same components in two swimming modes, *Exaggerated Swimming* and *Slow Swimming*. The first mode was characterized by wide-sweeping movements of the tail, which were necessary to propel the shark forward with the prey held in its mouth. A shark exhibited *Slow Swimming*, characterized by smaller movements of the tail, when traveling without the prey in the mouth.

It would be easy to measure the duration of a tail elevation and water splash. I simply counted the number of frames on the video-recording and multiplied this count by the time elapsed per frame. However, it would be a more daunting task to measure the amount of the shark's body held out of the water or the distance that water splashed after the tail contacted the water, yet I did this using an image analysis system in the laboratory. The resulting measurements indicated that the body movements in the *Tail Slap* were more exaggerated than swimming movements, with or without the prey in the shark's mouth. For example, a shark's tail was extended out of the water an average (in actuality, a median) distance of 100 centimeters during 63 *Tail Slaps* versus 70 centimeters for 31 periods of *Exaggerated Swimming* and 60 centimeters for 43 periods of *Slow Swimming*. As much as 180 centimeters (close to 6 feet) of the shark's body was elevated during a *Tail Slap*. The *Tail Slap* resulted in water splashing a median distance of 3.5 meters versus 2.5 meters for *Exaggerated Swimming* and 1.0 meter for *Slow Swimming*. In one instance, the shark propelled water 9.0 meters with a *Tail Slap*. That's a long way, much longer than the panga, my 23-foot research skiff.

Establishing the last prerequisite for communication—demonstrating that the intensity of the display of the winner in a tail-slapping contest was greater than that of the loser—was the most formidable challenge of the three. I needed to compute a mathematical value for the vigor of the display. It was not simply enough to compare the number of *Tail Slaps* by one shark to the number performed by the other and declare the one performing the greater number winner of the contest. The amount of the body out of the water during each *Tail Slap* was already known, and this varied from nearly half the shark's body, 185 centimeters, to just the upper part of its tail, 70 centimeters. It struck me that the sharks might judge the physical prowess of the other member of the contest from the degree of vigor in each of its each *Tail Slaps*. Surely the more of the body held out of the water by the shark, the more imposing it would look to its opponent, the louder the sound resulting from the tail hitting the water would be, and the more water would be splashed at the opponent. And as noted previously, the distance that the water traveled in the air also varied greatly, from as far as 9.0 meters to only 1.0 meter. It appeared, then, that all *Tail Slaps* could not be considered equal. What I needed was a way of expressing the relative signal strength of each *Tail Slap*!

At this time, Will Mangen, an electronic whiz, was working in my laboratory on developing an electronic tag that would determine its geoposition (its location on the earth) by inferring the time of sunset and sunrise from the associated changes in underwater illumination. He had read a book, called *Fuzzy Logic* that described a technique used to make decisions from multiple inputs of information. He intended to use this technique to better relate underwater changes in illumination to the position of the sun in the sky. At his suggestion, I bought the book and was reading it in the evenings before going to sleep. It struck me one sleepless night, thinking about sharks slapping their tails on the water, that this same technique could be used to obtain a quantitative index, or an indirect measure, of the display strength of the *Tail Slap*.

The word "fuzzy" refers to an alternative approach to classifying statements or information as two simple states such as true or false, yes or no, or 1 or 0. A property of what is measured is given a value in intensity ranging from a minimum of 0 to a maximum of 1. An everyday example of how this logic is applied to problems is as follows. A "fuzzy" washing machine would automatically determine the best wash-and-rinse cycle on the basis of information from several inputs. First, the weight (or amount) of clothing would be measured by a sensor attached to the receptacle containing the clothes. Based on the range of weights obtained for a variety of laundry loads, the sensor would be calibrated so that the heaviest weight gave a value of 1; the lightest, 0; and an average, 0.5. Second, the amount of dirt in the clothes would be detected by passing through the water a beam of illumination from a light-emitting diode to a light-sensing photocell. The amount of light that passed through the water would be inversely proportional to the amount of dirt dissolved in the water: the less light reaching the photocell, the more dirt is in the water. The sensor would then be calibrated so that the minimum light transmission was 1 and the maximum would 0. The machine would choose a particular rinse cycle based on the sum of the values for these sensors.

I took the four components of the *Tail Slap*—length of body out of water, duration of body elevation, splash distance, and splash duration—and treated them in a similar manner. The length of the tail lifted out of the water was measured for 63 of 83 *Tail Slaps*. These lengths varied from a minimum of 64.5 to a maximum of 183.3 centimeters. A nondimensional index of signal strength (termed SS) was calculated for each tail elevation, ranging from 0 to 1.

Cut Dorsal, another shark that often visited the island and was missing the tip of its top fin, elevated 170.2 and 181.6 of its body out of the water during the first and second *Tail Slaps* during an attack on September 30, 1989. The index values were 0.874 and 0.969. These were added to give a cumulative index of 1.843. The opponent of Cut Dorsal raised 139.2 centimeters of its body from the water only once and had an index value of only 0.616. Similar index values were calculated for each of the two sharks from measurements of the duration of tail elevation, splash distance, and splash duration. Each of the four components of the display were added to give a cumulative index of behavioral vigor for each of the two competitors. In predation P-104, the two *Tail Slaps* of Cut Caudal had a cumulative signal strength (see SS above) of 5.855, much greater than the 1.838 cumulative value for the single *Tail Slap* of its opponent. Cut Caudal was the winner!

I also developed an index of "feeding success," because it was not always possible to actually see which shark fed on the seal after the aggressive encounter. In addition to bites to the prey, two other indicators of feeding success were recorded, the amount of time spent at the surface near the prey and the frequency of visits to the surface. We calculated the same nondimensional values for the degree of success for each competitor in the same manner. The individual with the more vigorous *Tail Slaps* won the prey in fourteen of sixteen predatory attacks for which measurements could be made of signal strength and predatory success. Even when the value of the display for one shark exceeded that for the display of another by just the slightest amount, the winner appeared frequently at the surface to consume the prey while the loser almost never came to the surface in search of the prey. Finally, the winner of the combat was not always the larger shark, but at times was a smaller shark that displayed more vigorously. For example, Intact Caudal, who was 406 centimeters long, won the contest with Cut Caudal, who was 433 centimeters long, during attack P-191 on November 7, 1990, despite being almost 30 centimeters (approximately a foot) shorter than the shark missing part of its tail. It is possible that the sharks need to perform more *Tail Slaps* when the size discrepancy between individuals is small. In this attack, the two sharks passed by each other three times splashing water at each other. Intact Caudal performed fourteen *Tail Slaps* to six completed by Cut Caudal.

It was really difficult to convince others of the significance of a behavior that you have observed and they have not. While those of us at the Farallon Islands were alone in observing white sharks perform the *Tail Slap* in the context of predatory encounters with seals, others had observed the *Tail Slap* displayed by sharks attracted by bait to the side of a boat. In this case, the shark's tail may have been raised when it tried to submerge with a piece of meat in its jaws. Alteratively the shark may have performed a threat display because it confused the shark cage or boat with another shark for an opponent. With regard to a shark cage, I observed at Dangerous Reef, South Australia, that on two occasions the shark cage drifted between the shark and baits tied to lines off the stern of the boat. Both times, a shark elevated its tail out of the water and brought it down forcibly to strike the top of the cage.

Ken Goldman and Scot Anderson observed the same behavior from a small boat floating near a seal carcass on which two sharks were feeding. They dropped a piece of seal fat with a transmitter concealed within it into the water beside the boat. The objective was to get the shark to swallow the transmitter and then track the shark. Ken wrote in his diary what happened next: "When we arrived, no blood, no sharks, just gulls. Not for long: first one shark, then a second arrived. One shark bit the engine twice within 5–10 seconds. Then the boat was bumped on port side and tail slap shot water (two times) into boat (really across it). Got soaked good. Shortly thereafter (10–15 seconds), shark on starboard side bumped boat, bit engine, and tail slapped multiple times. Got completely soaked. It was incredible. Then another shark appeared. These sharks were 'hot' so we headed off." Scot Anderson's question to Ken at this time was a classic of understatement: "You still don't believe in the *Tail Slap*?"

As I explained afterward in our article on the *Tail Slap*, an alternative to a swift departure of Ken and Scot would have been to perform an experiment. The two, both recipients of the signal, could simulate *Tail Slaps* by contacting the water surface with an oar, thus propelling water in the direction of the displaying shark. I bet that if Ken and Scot had performed these artificial *Tail Slaps* more often and vigorously than the two white sharks performed their *Tail Slaps*, the sharks would have left the vicinity of the boat. But simulating *Tail Slaps* at that particular moment may have been over and beyond the call of scientific duty!

CHAPTER 12

BABY WHITE SHARK GETS AWAY
ON NATIONAL TELEVISION

"Pete, can you come down to Sea World today and track our baby white shark?" This was the second of two baby white sharks Sea World had recently kept in its shark aquarium in San Diego; one had died barely a month ago and the health of this one was rapidly declining. Soon it would have to be released into its environment or else it too would die in its tank. "I know this is short notice, but can you do it?" Jerry Goldsmith spoke in an apologetic tone as I listened intently on the telephone 400 miles away in my office at the Bodega Marine Laboratory, north of San Francisco. It was 9:30 A.M. on a Thursday, the first week of August 1994.

Jerry, a tall middle-aged man with blond hair, was the chief designer of new exhibits for the Sea World amusement parks across America. In a corner of his office next to the desk was a drafting table where he often sat for hours sketching plans for new exhibits that might someday become the home for dolphins, polar bears, sea otters, seals, or fishes. He and Mike Shaw, the head of the Aquarium Department at Sea World–San Diego, were keenly interested in keeping a white shark alive and healthy in their marine park for public display. They were not unique in their ambition, however—it was shared by the curator of fishes at almost every public aquarium in the world. The only obstacle to this universal desire of putting a white shark on display was keeping it alive—no one had yet kept one living in an aquarium for more than two weeks.

Jerry explained that when we had talked yesterday about releasing the juvenile white shark a week from now and tracking her, she was swimming normally like the other sharks in the tank. But her health had taken a sudden downturn overnight. Early this morning, they had noticed that she swam with her tail bending downward, and this forced her head and snout upward toward the surface, like a rocket faltering upon launch. She seemed to be laboring very hard to swim around the tank. She was now bumping into the walls as she circled the tank. Her snout was a reddish white because the dermal denticles, the toothlike scales on her body, had been scraped off when she made contact with the side of the tank, exposing the raw skin underneath. Jerry and Mike wanted to release her right away but were still committed to learning something about her behavior after releasing her.

Mike Shaw came on the line then and told me, "We want to let her go in the same place that we caught her—close to the shore at the northern base of the canyon." He was referring to the La Jolla Submarine Canyon, a massive underwater valley over 15 miles long, several hundred fathoms deep in the center, and flanked by steep sides, that led away from shore in a northwesterly direction and then bent toward the southwest. The canyon split in two close to shore; one arm originated 1 mile south of the pier of the Scripps Institution of Oceanography and the other a quarter of a mile north of the pier. The place was familiar to me—I had swum in those waters while a graduate student at Scripps, observing members of a large school of leopard sharks forming early in the spring near the base of the canyon in the shallow water off a cluster of rocks. Mike then asked, "Do you have a spare transmitter that would tell us something about her swimming behavior and where she goes once we let her go?"

I was impressed with Jerry and Mike's commitment to learn more about the behavior of the little shark—this kind of knowledge would be vital to keeping one of her kind alive in a public display. I have always been supportive of the ambition of many professional aquarists to keep white sharks in captivity—having kept tropical fish in aquariums for most of my life—but only if the tank could simulate the essential properties of the shark's natural habitat. It would obviously not be possible to make a tank with the same dimensions of the white shark's oceanic habitat. One could, however, build an artificial habitat that duplicated only those factors critical to the species' healthy existence. The tank would have to be sufficiently wide to permit the sharks to accelerate with several strokes of the tail and glide before having to reverse direction, and the water temperatures would have to be the same as in its natural habitat.

The design of this public display would have to be based on sound knowledge of the species' behavior and physiology, and the best source of this information would be observations of a white shark in its normal environment. I had for a long time been working with Jerry to interest Sea World in financing a study in which juveniles with ultrasonic transmitters would be tracked in their own environment. The transmitters would be equipped with sensors to detect behavioral characteristics of the shark such as its swimming speed and depth, and environmental properties such as water temperature and illumination. It was my dream, as well as theirs, to build a white shark–friendly exhibit that succeeded as a home for the species. The public would not be solely dependent upon viewing film documentaries to learn about the species, but could see white sharks close up and better understand the nature of this awesome predator.

I told Mike that there was a spare transmitter in my laboratory that we could use to track her. We would also need a boat equipped with navigation gear so her position could be recorded while tracking her. We could then determine her rates of movement and degree of orientation between successive positions. However, we would not be able to record her swimming depth and the water temperature because there was no pressure gauge or thermistor on the device. It was my hope that my joining Sea World in this spontaneous venture now would pave the way for a joint, carefully planned study of the juveniles in the future using transmitters with these sophisticated sensors. But for now, the beacon would tell us something about their shark's rate of movement, where the shark went, and, most important, whether she lived after being released. The public would surely want to know that last item of information. I had already been contacted by both radio and television newspeople concerned about the health of the captive shark.

I told Jerry and Mike that I would fly down to San Diego immediately. There would be no time to stop by my house and pick up extra clothing. It would take two hours to drive to the San Francisco airport from Bodega Bay, another hour to fly from San Francisco to San Diego, and a half hour to go by taxi from the airport to Sea World. That would make three and a half hours until my arrival at Sea World. It was now ten in the morning—I would be there by two-thirty in the afternoon.

The drive to the airport and the flight down to San Diego went smoothly. After getting off the plane, I rushed out of the terminal with a pole spear in one hand and the suitcase with the receiver in the other and briskly walked to the curb where a taxi was waiting to take me to Sea World. The taxi took off with a screech, with me in the back, a little apprehensive about the likelihood of my arriving at Sea World alive when the driver was speeding so fast. Arriving at Sea World, I trotted up to the employee entrance. There waiting for me was Pamela Yocum, an extremely talented person who divided her time between being a researcher at the Hubbs-Sea World Research Institute, a research unit associated with the aquarium, and a veterinarian at Sea World. "Aha," she said. "You are Dr. Klimley, the world-renowned shark expert—just the person we need." This was a surprise—I wasn't accustomed to such flattering introductions. She said, "Come with me and I will lead you to the boat in which we are going to take the shark out to the submarine canyon."

She introduced me to person after person as we briskly walked through Sea World toward the dock. One of the people I spoke with was another veterinarian, Tom Reidarson. He had taken blood from the two baby sharks before they were released into the aquarium and again prior to the first shark's death and this shark's release. He informed me that the glucose (sugar) and insulin levels in the first shark were very high prior to its death—this hyperglycemia was probably a symptom of stress in the aquarium environment. What impressed (and worried) me was how grateful everyone was that I had come down from Bodega Bay to track the shark. They seemed to expect me to pull off some kind of minor miracle, and I worried whether my performance would meet their high expectations.

As I walked through the alleys behind the public exhibits toward the dock where the boat that would transport the shark was tied, I noticed on the other side of a chain-link fence out in the park a group of people milling around, some of them carrying signs. It seemed like they were shouting slogans. I was informed that these were animal rights activists, alarmed over the possibility that another baby white shark would die in captivity. The shark kept at Sea World during July had died after less than two weeks in its exhibit. These people were determined that this baby shark would not suffer the same fate.

The boat for transporting the shark had a wide-open space in the bow for the shark's enclosure. I stepped into the stern of the boat, opened the suitcase, and began to unpack the telemetry gear. Standing to my side was the official photographer of Sea World videotaping everything that I was doing, coverage obviously intended to be used for public relations. I took the hydrophone and additional sections of metal rod from the suitcase and started screwing them together to make a long staff with the hydrophone at one end and the handle at the other. When this was completed, I bent down, took the plug from the hydrophone, and pressed it into the back of the cylindrical receiver. I put the earphones on my head and turned on the receiver. Listening intently, I tapped the rubber face of the hydrophone twice with my forefinger and smiled as I heard a *bink-bink*, indicating that the receiver was working.

Now I had to make sure that the transmitter was working. I picked up the small device, the size of a small test tube, and removed a little magnet taped on its side. I activated a transmitter using a magnetic reed switch. This was a miniature glass ampoule with two metal plates passing through either end and continuing inside until they came in contact, one slightly overlapping the other. When a magnet was taped to the transmitter above the switch, its repellent force pushed one of the plates away from the other, keeping the switch in an off position. Remove the magnet, and the current would flow and the device would emit its high-pitched pings. I held the transmitter next to the hydrophone at the end of the metal staff, listening to the loud *bink-bink-bink* over the earphones that verified that the transmitter functioned. I then taped the magnet to the transmitter, turning it off. I next placed the device on a flat plastic sleeve near the end of the pole spear, put a rubber band around it to keep it in place, and inserted the sharp dart attached to the transmitter into the slot in the applicator tip at the end of the spear. I gently laid the spear and transmitter along the side of the boat. I then neatly placed the receiver and the earphones in the foam compartments for them in the suitcase. I was ready to go!

Appearing in the distance and slowly coming toward the pier was a large rectangular box, suspended from a metal railing and surrounded by people wearing Sea World jackets. This must be the life-support system with the baby white shark! As soon as the box was alongside the boat, it was moved sideways and lowered into the boat. There, lying in clear water, was a beautiful miniature white shark, 5 feet long. This was the first time I had been able to see a baby white shark up close.

She was light gray on top and white below, not the slate gray and white of an adult. I thought to myself that less light would illuminate her white undersides, shaded from the sun, than her darker back. She would thus appear a uniform gray underwater and would be hard to see from the side, unlike an adult with a sharp boundary between the dark back and white underside. That indicated that she probably captured her prey in midwater or just off the bottom, not near the surface like the adults. Another impressive feature was her caudal peduncle. This was the muscular connection between her body and tail, which was comprised of a large upper lobe and a slightly smaller lower lobe. The peduncle was huge and had outward-projecting keels on either side. It must be close to 15 centimeters (6 inches) in diameter—a lot of muscle for such a small shark! I thought to myself that not only would she be hard to see underwater, but she must be a magnificent swimmer, the perfect combination of traits to enable her to chase and capture small fishes that schooled along the edge of the submarine canyon. I looked at her mouth and could see her many rows of pin-sharp pointed teeth for seizing and swallowing fish once she chased them down. This juvenile was quite different from the adult white shark! The adult's dark gray dorsum, or back, matched the mottled grays of a rocky bottom and enabled it to swim along the bottom until it was underneath its prey. Then it would dash toward the surface to ambush an unsuspecting seal or sea lion. The adult had large triangular teeth that were serrated, ideal for sawing off large chunks of meat from adult pinnipeds. In fact, an adult white shark could easily sever the very dense bone of the vertebral column (almost granitelike in its weight) to decapitate a seal or sea lion, often the first act upon capturing one.

Mike's assistant, a top-notch aquarist in his own right, introduced himself to me. Pointing at the transmitter near the end of the spear, I said that the dart in the spear's tip would be inserted into the thick musculature on the shark's back just before release. I would then take the metal staff with the hydrophone at one end and place it in the water, determining the direction to the shark by rotating it back and forth until the signal was strongest. Then we would drive off in the boat after the shark. He was standing proudly in the front of the tank with the baby white shark, and pointed out to me the canvas sling that passed underneath her and was attached to two pipes on either side of the bottom of the container. He and the person in the back would then reach down into the container, grasp the front and back handles of the sling, and lift the shark, weighing close to 200 pounds, up and out of the container. They would then swing the sling over the side of the boat and lower it under the water so that the shark could swim off to her freedom.

Once this was explained, another member of the Aquarium Department started the boat's engine, our lines were cast off the dock, and we began our journey toward the submarine canyon. As we drove out of the Sea World Marina, converging upon us from every direction were boats crowded with spectators who wanted to witness the release of this celebrity. Boats lined up on either side of us and behind in a massive procession that left Mission Bay and slowly snaked along the coast of La Jolla toward the Scripps Pier. When we reached the mouth of the bay, there was a loud and rhythmical *whompah-whompah-whompah*, and we all looked upward and saw a helicopter hovering directly over us. Hanging out of the side door of the helicopter and filming us was somebody from a local television station. Gazing at the cameraman and then glancing at the Sea World photographer next to me, I felt apprehensive and hoped that, given all of this media attention, everything would go well when we released our celebrity.

It took about an hour before the procession of boats arrived at the submarine canyon a quarter mile north of the Scripps Pier. I promptly attached the transmitter to the shark. I then put the earphones on momentarily to make sure the receiver was working. We were now ready to release the shark. The two aquarists reached into the rectangular enclosure to get ahold of the handles on either end of the sling. As the senior aquarist's hands passed in front of the shark, it suddenly opened its mouth wide and projected its upper and lower jaws, each bristling with rows of pointed teeth, outward and closed them on one of his hands. "Oooowch!" he shouted. Yet in spite of the pain, he managed to grab the snout of the baby shark with his free hand and lift it upward to open the shark's mouth and quickly draw his other hand from the shark's mouth.

He and his assistant then lifted the sling and shark upward, moved it over the side of the boat, and lowered it into the water until the shark was entirely submerged. Suddenly, there was a large splash at one end of the sling as the shark swept its tail back and forth and accelerated like a torpedo out of the sling into the ocean. Once out of the sling, she accelerated out of sight, beating her tail back and forth energetically, not at all like her slow and labored swimming in the aquarium. She appeared to be perfectly healthy in her normal environment.

I quickly lifted the hydrophone staff, put it in the water, and placed the earphones on my head. As I rotated the sound-detecting element away from shore, the *bink-bink-bink* became louder. The shark was heading offshore. I rotated the gain control clockwise to increase the receiver's sensitivity to the signal, and listened through the earphones for the transmitter. The pulses were loud, indicating that the shark was not that far from us, but were decreasing in volume because the baby white shark was rapidly distancing herself from us. We had better take off immediately in pursuit of the shark and drive the boat fast for five to ten minutes to keep up with it. I pointed offshore and said, "There, drive the boat in that direction."

I then had a second thought—it was surely more important to make sure the aquarist's hand was OK. I had assumed his hand was not harmed because he had used it to lift the sling out of the boat and lower it into the water to release the shark. My eyes then immediately shifted in the direction of the senior aquarist, who had been "attacked" by the baby shark. He was holding his hand, which by now was covered with blood seeping from puncture wounds inflicted by the sharp teeth of the baby shark. It stuck me now that our first priority could not be to track the shark, but must be to administer first aid to his hand, even if that were to mean losing the shark. We quickly drove the skiff over to a bigger boat with a cabin not far from us. This boat would be used to track the shark at night. Standing in it was Tom Reidarson, the Sea World veterinarian. He held the aquarist's bleeding hand in one of his hands, looked at it closely, and used his other hand to point out a series of puncture wounds inflicted by the shark's pointed teeth. He told us that the aquarist had been very lucky. First, the shark had not bitten down that hard on his hand. Second, the aquarist had the presence of mind to not jerk his hand out of the shark's jaws, an act that would have resulted in the teeth tearing his flesh. He had opened the shark's mouth before withdrawing his hand.

The aquarist needed to be taken back to Sea World immediately for medical treatment. He and Tom boarded the small boat and were driven back to the aquarium. The Sea World photographer and I remained on the larger boat. The skipper promptly drove this boat back to where we had released the shark, and I placed the hydrophone staff in the water and began rotating the staff back and forth, searching for the signal from the shark's transmitter. To my despair, all that was audible was the high-pitched crackling sounds of snapping shrimps. In desperation, I rotated the knob clockwise as far as it would go to select maximum gain, but instead of hearing the faint pulses of the transmitter, I heard only the shrimp.

We couldn't give up yet when there was still a chance that we could relocate the shark. I told the skipper to drive the boat five minutes in an offshore direction toward the northwest—the shark had taken off in that direction twenty minutes ago. I again placed the hydrophone staff in the water and listened for the signal of the transmitter. There was no *bink-bink-bink*. I now told the skipper to continue in that direction for ten minutes. Tagged sharks could often be relocated in this manner, but only after they had been tracked for some time and their general direction of movement was known. We had not yet tracked the baby white shark at all, aborting the track to take care of the aquarist's hand. I again put the hydrophone in the water and listened. This time, I looked up at the Sea World photographer, who was standing there videotaping me, and then looked up at the sky at the cameraman in the helicopter, who was also taping me. I then thought to myself, "———! I've lost the shark, and practically everyone in the country has witnessed me lose it on national TV."

We didn't give up then, but spent another three hours searching for the shark in the waters off the coast of La Jolla, giving up only after the sun had set and it had become dark. I was one truly exhausted marine biologist that night, falling immediately asleep in the guest room of Jerry Goldsmith's home. However, I rose early the next morning, drove over to Sea World with Jerry, and accompanied several members of the Aquarium Department as we searched for the shark until noon. We had no luck locating her—one must realize that it is a true challenge to find a single fish in the vast ocean. Upon returning to Sea World, Jerry informed me that there would be a press conference at 1:30 P.M. at which reporters from several radio and TV stations would ask me what happened to the shark.

That afternoon, reporters from the local newspaper, radio, and television stations asked me questions about the baby white shark while standing in front of the shark exhibit. I was asked not to mention the name of the person whose hand had been bitten by the shark (and his name is not included here). It was not easy to keep this secret when it contributed to our losing the shark, but I agreed not to mention the injury unless asked specifically about it. Later, the video depicting what had happened that day was shown to other aquarists at a national meeting and the "shark attack" became public knowledge.

I emphasized at the press conference that although we had failed to track the white shark this time, any future tracking would be better organized and would have a better chance of success. I wanted to put the best face on things because Sea World had agreed to collaborate in a tracking study aimed at describing the behavior of the species in its natural habitat in order to design any future aquarium, one that would meet the shark's physiological needs. I left San Diego for San Francisco late that day, looking forward to working again with the members of the Aquarium Department. As I passed through the baggage checkpoint at the airport, several of the employees recognized me. One said exuberantly, "Oh, the shark got away and is well. I am so happy." The others clapped their hands in joy.

The last day of September, I met with Jerry, Mike, Tom, and others at Sea World to plan our future research activities. We decided to fish for juvenile white sharks only during the summer months, when members of the Aquarium Department were least busy. One or more juveniles, carrying transmitters with sensors recording water temperature and swimming depth, would be tracked for one to three days. During each track, we would periodically lower a bathythermograph, an electronic device that stored a record of temperatures at various depths. The purpose for measuring the variation in water temperature with depth was to ascertain whether the shark swam only in water that stayed in a narrow range of temperatures. The temperature of the water in an exhibit could then be kept within this optimal range.

Members of the Aquarium Department would fish for white sharks using their two motorboats. Each aquarist would take another person's place caring for his or her aquarium for the day. Later during the study, Sea World hired additional people to assist in the fishing activities for the sharks. Each boat would set approximately twenty fishing stations in water 40 to 100 feet deep along the edge of the submarine canyon. Each station would consist of a small anchor attached to a line leading to a buoy on the surface with three or four leaders with hooks at various lengths. The fishermen were to bait the hooks, drive the boat some distance from the stations, wait while drifting for two hours, and then return to examine what had been caught by the stations. Moorings were used for fishing stations instead of a long line with many leaders and hooks, because once it was hooked, the shark would be able to swim around the mooring and force oxygenated water past its gills in order to breathe. Each boat would carry a transport box to hold the white shark once it was caught by one of the fishing stations.

The *Pelican*, the research vessel of the Hubb-Sea World Research Institute, a research organization associated with the aquarium, would be used for tracking the shark. This was a moderate-sized fiberglass motorboat with a cabin. Outfitted with tracking gear, it would be on continual standby with a crew consisting of members of the Aquarium Department ready to track the shark immediately upon notification by one of the fishing boats. Once captured, the white shark would be lifted into the transport box, which contained water that was continually circulated with a small pump and oxygenated with a stream of bubbles of oxygen released from a cylinder. The aquarists would take a small sample of blood, to ascertain the stress level of the shark, and a biopsy of muscle for genetic analysis, which could tell whether the sharks caught were family members. The transmitter would then be attached to the shark. It would be released as soon as the *Pelican* arrived and the tracking crew was ready. I felt that the members of the Aquarium Department were capable of tracking the sharks themselves. I agreed to come to Sea World the following June in order to install the tracking system on the *Pelican* and train the aquarists by tagging and briefly tracking two blue sharks in the La Jolla Submarine Canyon. They would then fish for white sharks during July and August.

I arrived at Sea World during the second week of June 1995. My first task was to install the tracking equipment in the *Pelican*. The metal shop at Sea World built a bracket to attach the hydrophone staff to the side of the boat, where it could be rotated to locate the direction to the shark. I connected the receiver to two laptop computers, one that would provide a continuous record of a shark's geographical coordinates, rate of movement, swimming depth, and surrounding water temperature and another that would plot the position of the shark periodically on a nautical chart displayed on the screen. The shark's position would be estimated by slowly driving the boat over the shark and acquiring the boat's geographical coordinates with a

GPS receiver connected to the second computer. At this time, the military purposely reduced the accuracy of GPS coordinates to prevent hostile countries from using the technology to guide their weapons to targets. The magnitude of the spatial error for the positions of the shark was illustrated one day when we drove the boat out of Mission Bay. The channel leading to the ocean was bordered by seawalls composed of large boulders piled on top of each other and was 40 yards across. The navigation program displayed our position on an electronic chart. The boat was not shown moving down the center of the channel, but on the land on the other side of one of the seawalls. I immediately added a differential receiver to the system that detected a signal from a radio beacon situated on land and, based on this, eliminated the introduced error in the satellite signals.

We conducted two practice tracks of blue sharks on the eighth and the fourteenth of July. Both sharks were caught a mile offshore of Sea World just after noon and tracked for two hours. The sharks were caught with a fishing rod after being attracted by ground fish we had put into the water. The blue shark was kept in the transport box until the Pelican arrived with half a dozen members of the Aquarium Department. The shark was then tagged and released in a similar way as the baby white shark. Each of the aquarists took his or her turn, rotating the hydrophone, listening for the signal of the transmitter, and then indicating the direction of the loudest signal to the person at the helm, who then drove the boat toward the shark. Blue sharks are easy to track. The two blue sharks swam very slowly, 0.3 to 0.6 meters per second—a human walking briskly moves about 1 meter (3 feet 4 inches) per second. A transmitter with a pressure sensor providing a record of depth was attached to the second shark, which made frequent but shallow dives, continuously moving up and down in an oscillatory manner. Both sharks slowly moved offshore in a straight-line manner. I returned to the Bodega Marine Laboratory at the end of the week, satisfied that the Sea World aquarists were well prepared to track a white shark.

Only four days later, I was alerted by Sea World late in the afternoon that they had caught a baby white shark and were tracking it. The small male shark, 152 centimeters long (nearly 5 feet) was tagged and then released a mile north of the submarine canyon. It briefly swam at the surface—the aquarists standing on the deck of the boat could see it—less than 25 yards from shore within the surf zone. The baby shark meandered by several body surfers, who were standing in the water, oblivious to its presence, waiting to ride the next large wave toward shore. The shark then turned offshore in a northwesterly direction and swam along the edge of the submarine canyon. The white shark moved at a rate of 0.8 meters per second, exceeding the rate of movement of both blue sharks. The faster speed may be due to the white shark's elevated body temperature relative to the blue shark. At a higher body temperature, more oxygen can bind with the hemoglobin in the blood and be transported to the muscles of the tail, permitting it to beat faster.

When close to shore in the shallow water, the shark alternated between swimming near the bottom and on the surface, but once it reached deeper water, it made oscillating yo-yo dives between the surface and a depth of 20 meters (66 feet), which was far from the bottom. The aquarists lost contact with the white shark at 6:36 P.M., when another boat from Sea World arrived with a second team of aquarists to track the shark during the night. The engine sounds from the approaching boat may have frightened the baby shark, which unexpectedly accelerated out of the range of reception. Yet it was very gratifying to everyone to have tracked a white shark after fishing only four days at the edge of the submarine canyon. However, little did we know at that time that we would not catch another white shark during the rest of this summer, as well as all of the following summer, despite two crews fishing every day at the edge of the submarine canyon.

Having failed to catch another white shark during 1996, we decided to track three mako sharks during the following summer. The rationale for this decision was that because the mako and white were members of the same family, the mackerel sharks, they might share common behaviors and physiological requirements.

There are close to four hundred species of sharks in the world, a number that is changing as new species are discovered. The most recent discovery was the 13-foot-long black megamouth shark, which was placed in a new family. This is a large, slow-moving, planktivorous shark with a cavernous mouth and gill rakers with many little projections for collecting plankton. The megamouth, which lives in the open ocean, forages in the deep water during the day and likely rises only to within 150 meters of the surface at night to feed on deep-sea euphausids. These bioluminescent shrimplike animals are vertical migrators, which spend days in deep water (300–1,100 meters) and nights closer to the surface (150–300 meters).

Sharks are in the class Chondrichthyes, and possess cartilaginous skeletons, composed of soft and rubbery calcium phosphate, in contrast to the bony fishes, whose skeletons are formed of hard and stiff calcium carbonate. Within this class, there are eight orders of sharks, most having cylindrical trunks to their bodies, and one order of skates and rays with flattened bodies. The eight orders of sharks are the angel sharks, saw sharks, dogfish, frilled and cow sharks, horn sharks, nurse and carpet sharks, ground, or reef, sharks, and mackerel sharks. The members of the first three orders do not have an anal fin, a small bottom fin immediately in front of the caudal fin; members of the remaining five orders do have this fin.

First, a little about the three orders of sharks without anal fins. The angel sharks are small species (less than 5 feet long), that are raylike, with a flattened body and a mouth at the front of the body. These lie buried in the mud and sand, where their highly protrusible, traplike jaws, can suddenly open, creating a vacuum and drawing in unsuspecting fishes and crustaceans swimming over them. The saw sharks, on the other hand, possess a snout elongated into a sawlike blade with teeth protruding laterally from the edge of the rostrum on either side and a pair of long barbels protruding from its underside. These small (less than 5 feet long) and very slender species are thought to move along the bottom, detecting their prey by vibration or electrical sensors on the underside of the rostrum and killing them by striking them with their rostrum. The dogfish are cylindrical sharks, varying in size from the small cookie cutter, less than 3 feet long and possessing razor-sharp teeth in the lower jaw that are twisted to scoop out a conical plug of flesh from large fishes, seals, and dolphins, to the sluggish sleeper shark, reaching 23 feet long and at times capturing its prey by opening its mouth wide to create a suction (like a slurp gun) and drawing a seal inside, while twisting to remove its outer fatty layer. The most common species in this order are the dog sharks, and some of these form immense schools that are highly nomadic—and this similarity with wild hunting dogs has led to the group's name. Many members of this order inhabit the deep sea.

There are five orders of sharks with anal fins. The frilled and cow sharks are small-to-large sharks (some reaching 15 feet long) that differ from all other species of sharks in having one dorsal fin and six or seven pairs of gill slits. Most shark species either swallow or force water through their mouths and across their gills, extracting oxygen, and vent it through five gill slits. The horn sharks are small sharks (most are less than 4 feet long) with a piglike snout and two dorsal fins that are armed with sharp spines. These species have highly specialized jaws, with small, sharp, pointed teeth at the front of their jaw for grasping sea urchins, crabs, and clams. These teeth are used to seize prey and force them backward into the back of the jaw, where pavements of heavy molarlike teeth on both the upper and lower jaws crush the outer tests, carapaces, or shells of their prey. The nurse and carpet sharks are medium-sized sharks with their mouths far forward of their eyes and large barbels extending from the inner edges of their nostrils. These species are often flattened, cryptically colored with variegated color patterns consisting of bands and large spots, and have mustachelike dermal lobes around their chins. These species often remain on the bottom during the day, waiting to ambush prey swimming above them, but also actively search at night for crabs, shrimps, squid, and small fish. Huey, the fast-learning subject of my experiments on shark intelligence, was a member of this order. The whale shark is also a member of this order, although its planktonic feeding mode differs from that of the rest of the species in the order.

The ground, or reef, sharks comprise the largest of the orders of sharks. They can be distinguished from the nurse and carpet sharks by the fact that their mouths are behind the front of their eyes. They have nictitating membranes, which are eyelids that shut from below, and either spiral or scrolled valves to increase the absorptive area in their intestines. The leopard shark, which forms schools close to shore off Southern and Central California, and the lemon shark, the subject of my killer-whale-suit experiment, are members of this family, which live in coastal waters; other members of this family include blue, silky, and whitetip sharks, which inhabit deep water in both the Atlantic and Pacific Oceans. The hammerhead sharks differ from the other members of the order in the lateral elongation of their rostrum into the shape of a double-bitted ax or mallet—the hammer-shaped head. The size of this rostrum varies among these species. The winghead, which lives in the Indian and Pacific Oceans from the Persian Gulf to the eastern coast of Australia, has protruding, winglike extensions to its rostrum, whose width is close to half its body length; the bonnet-head, which inhabits the tropical waters on the eastern and western shores of North America, has a small, rounded rostrum. The scalloped hammerhead, the subject of my behavioral studies in the Gulf of California, is the most widely distributed and populous species in this family.

The mackerel sharks differ from their nearest relative, the ground, or reef, sharks, in that they possess no nictitating eyelids and have a ring valve instead of the spiral and scroll intestinal valves of the ground sharks. This is the order that contains the "money" shark, the one that sells novels, movies, documentaries—the great white shark. Members of the order such as the mako, salmon, and white sharks are large, with pointed snouts, long pectoral fins and high dorsal fins, and possess two large lateral keels on the side and pits on top of their caudal peduncles (the narrow connection of the body to the tail). The mako shark inhabits semitropical to tropical waters, the white shark temperate to tropical waters, and the salmon and porbeagle sharks (the former lives in the Atlantic, the latter in the Pacific) live in temperate to polar waters.

After reading my description of the range of their lifestyles, one might wonder if any similarities exist in the behavior and physiology among species of sharks. Actually, there are similarities among the species. To start with, the bodies of the mako and white sharks (both mackerel sharks) are warmer than the surrounding water because they possess retes (interwoven arteries and veins) that prevent heat loss. Heat passes from the warm blood in the blood vessels flowing away from the heart to the errartis exterior body to the cooler blood flowing toward the interior body and back to the heart. Hence, less heat is lost to surrounding environment. The retes act as a heat-exchange mechanism to warm the brain and musculature, increasing the transfer of oxygen to the tissues and quickening the reaction times of the nervous system in the cold waters inhabited by the sharks. Other species such as the blue shark (in the order of ground, or reef, sharks) lack these retes, and consequently their body temperature varies little from the water surrounding them.

I helped the aquarists from Sea World track the first mako, a female 119 centimeters (4 feet) long. She was tagged and released just after noon on June 25 over the northern slope of the submarine canyon and was tracked for twelve hours. Initially, she swam offshore in a northwesterly direction, but after two hours changed her course to the southwest. She swam along the edge of the canyon, but her alteration in course coincided with a change in the direction of the canyon, now directed toward the southwest and leading into the deep water off the continental shelf. She was swimming far from the bottom and surely couldn't see the canyon and use it as a navigation aid.

At dusk, I grabbed the hydrophone and began tracking the mako because we began to lose her outside the range of the receiver. Initially, she swam ahead of us, but then reversed her direction and ended up behind us. She seemed to be accelerating with bursts of speed, one reaching 7.8 meters per second (15 knots, or nautical miles, per hour), over seven times her median rate of swimming which was 0.9 meters per second (a pace equivalent to a brisk walk). I thought at the time that she was chasing small fish, which are often vulnerable at dusk, as they disperse from their daytime schools and seek their nighttime abodes alone. The mako has multiple rows of pointed teeth on both the upper and lower jaws, ideal for seizing a fish and drawing the prey back into its mouth to swallow it. The species' crescent tail, consisting of large lower and upper lobes attached to the body by a narrow peduncle, is suited for sustaining high speeds over distances up to a quarter of a mile. In its body anatomy and feeding mode, this species more closely resembles the juvenile white shark than the adult white shark.

During the rest of the day, she swam in a straight-line manner like the little white and blue sharks, first toward the northwest and then southwest. Her diving behavior also resembled that of the other two sharks. She performed the "yo-yo" swimming oscillations and at times swam on the surface of the ocean. Upon release, she made a deep dive that may have been caused by the stress of release, but after an hour rose to a depth of 10 meters (33 feet) and made regular dive oscillations between 5 and 20 meters, except for when she swam on the surface. She experienced a broad range of temperatures, from 50° F to 72° F. She was tracked over a distance of 22 miles before we lost her early in the morning.

During the summer, we tracked two more mako sharks for periods of twenty-two and thirty-eight hours. They exhibited the same behaviors and swam in water of a similar range of temperatures.

The behavior and movements of the different species of shark have usually been described within different geographical regions. These often differ in environmental properties—for example, there are cool waters in the temperate zones or warm waters in the tropics. Each study has focused on members of a single species. Nobody had compared the response of more than one species to the same environment until we did so with blue, mako, and white sharks within the submarine canyon off La Jolla. We wanted to identify any similarities in behavior among these three species, which might result from their using similar physiological mechanisms to cope with the same environment. The species had three behaviors in common. First, individuals of all three species swam in a directional manner. Second, they constantly moved up and down in the water column, exhibiting oscillatory or yo-yo swimming. Third, they swam at the surface for prolonged periods.

I had already observed impressive directional swimming in the nocturnal hammerhead migrations at the Espíritu Santo Seamount. Oscillatory swimming is one of the most common behaviors of large marine animals in the open ocean—bony fishes, cartilaginous sharks, seals, sea lions, toothed dolphins, and baleen whales all yo-yo in the ocean. Many reasons have been proposed for this universal behavior.

The reason most often proposed is to warm the body after heat loss during descent into cooler water. Tunas, makos, and white sharks maintain an internal body temperature above that of the surrounding water and, in this way, improve their muscular efficiency and enable themselves to swim at burst speeds. In their search for prey in the cold, deep waters, members of these species would need to surface periodically in order to reheat their bodies before returning to the colder water to actively pursue and capture prey. Consistent with this idea is the slow cooling of the muscles of tuna during descent and the more rapid warming of the muscles during ascent. It is as though during descent the tunas increase blood flow through their internal retes, forcing heat from the outward-flowing warm arterial blood to the inward-flowing colder blood in the veins, to avoid heat loss to the environment. However, the same tunas seemed to let the blood bypass the retes to warm itself during ascent. The change in the extent of oscillations between temperate and tropical waters also suggests that the range of the excursions is dependent on water temperature. The vertical excursions of our three sharks generally ranged between the surface and 50 meters and water temperatures of 75° F and 57° F. However, a mako shark tracked off Florida, where the same temperature gradient was distributed over a depth range of 400 meters, descended into much deeper water. Thermoregulation (regulation of the body temperature) is probably not the only function of oscillatory swimming, because it is also observed in cold-body fishes such as the blue and hammerhead sharks, which do not have retes to keep heat within their bodies.

Another proposed reason for yo-yo swimming is to explore the water column for directional information needed to migrate home. The oceanic water column is a collection of strata, each layer originating from a specific location and containing a unique chemical composition. As any fish descends, it passes through thick layers of water with the same temperature separated by thin layers with a large gradient (change) in temperature. Hakan Westerberg, a top-notch salmon researcher, argued that members of these species could find their home direction by moving back and forth between the stratum with an odor that came from the salmon's home waters and an adjacent stratum that did not have the distinctive odor. Of course, the salmon would need to use one of its senses to perceive the flow direction of the home water mass. A salmon might use its magnetic sense to detect the difference between the magnitude of electrical fields induced by the two masses of water moving in different directions. Alternatively, the salmon could use eyesight to distinguish between the different directions in which particles move in the two water masses or its contact-sensitive lateral line to detect the change in the pressure as water would flow in two different directions.

Westerberg supported his theory with two observations. First, the salmon that he tracked made wider excursions through the water column when their nasal passages were blocked than when unblocked, presumably searching vainly for the boundary between the source and the adjacent strata. Second, he and his colleagues found that nerves in the part of the brain devoted to chemical sensing responded more actively to water from particular strata along the migratory path of salmon.

A fourth reason for yo-yo swimming could be to minimize the expenditure of energy during swimming. It has often been suggested that the most efficient mode of travel for fishes is to alternatively swim upward with a certain number of tail beats and slowly glide down with fewer tail beats. Certainly, the shape of some diving oscillations support this fly-and-glide style of locomotion, which we also observe in birds. For example, take yo-yo oscillations of yellowfin tunas tracked at fish aggregating buoys off Hawaii. The upward slope of their ascent is usually steeper than the downward slope of their descent.

A fifth reason for oscillatory diving could be to better detect patterns in the magnetization in the seafloor—the navigation aid for hammerheads swimming over long distances. I have suggested that hammerheads dive in order to more easily detect the subtle magnetic topography of the ocean floor, which they use to guide their nightly migrations away from seamounts in search of forage. The hammerhead could distinguish a local gradient from the powerful main field (between the north and south poles) by gliding downward until the field rotates and increases enough for the anomaly to be perceptible above the main field. The individual would have to rise periodically to reestablish its field of reference. There is some evidence supporting this idea. Hammerhead sharks performed yo-yo diving oscillations as they swam along magnetic minima (valleys) and maxima (ridges) leading away from the Espíritu Santo Seamount like the spokes of a wagon wheel. I was able to show that during its highly oriented feeding migration, one hammerhead dove to that depth where the magnetic gradient was strongest across the magnetic lineation leading toward Espíritu Santo. It was as if the shark was seeking out the strongest magnetic gradient.

Surface swimming is also very common among migratory species. The behavior has been observed not only in cartilaginous fishes such as the blue, hammerhead, mako, and white sharks, but also in bony fishes such as the blue marlin, salmon, and yellowfin tuna. It is at the surface that it is easiest to use the earth's main field as a reference— the intensity gradient between the north and south poles is most uniform there. The hammerhead could swim in a constant heading either by keeping constant the induced field perceived by its receptors or by maintaining the differential between the field detected by its two bilaterally separated networks of receptors one on either side of the head. Another impetus for swimming at the surface would be to use the sun or moon as a directional reference.

In the future, we need to discriminate between these intriguing explanations of the function of the common behaviors observed among the species of sharks tracked in the submarine canyon off La Jolla. For example, one could perform an experiment to verify the connection between thermoregulation and oscillatory swimming. One could pass an electrical current through a low-value resistor warming that portion of the brain that turns on the retes while the shark is descending into cold water—one would predict that it would stay down longer during its dive. Or conversely, one could cool the same portion of the brain with a chemical coolant while swimming at the surface—one would predict that the shark would refrain from diving.

It is by monitoring the behavior of sharks in their own habitat and conducting experiments in the field as well as the aquarium environment that we will acquire the knowledge necessary to keep species such as the white shark in captivity. A full understanding of shark behavior requires not only open ocean observation but close-up observation over a long duration, which is possible only in a man-made environment. And we are a long way from knowing as much as we need to know about these—creatures, which may give us a unique, valuable perspective on marine life. But until we can find a way to maintain creatures like the great white in a controlled environment, we must seek them out in the oceans, an environment that can be just as challenging in its way as outer space.

CHAPTER 13

ELECTRONIC MONITORING OF WHITE SHARKS AT AÑO NUEVO ISLAND

The Animal Behavior Program of the National Science Foundation awarded Burney Le Boeuf, a seal behaviorist at the University of California at Santa Cruz, and me an eighteen-month grant during the spring of 1997 to study the predator-prey relationships between sharks and seals at Año Nuevo Island. My earlier five-year work in the Farallon Islands was a particularly good preparation for this study. We proposed to use a radioacoustic positioning system (abbreviated as the RAP) to monitor the predatory tactics of the sharks and evasive behavior of the seals within the high-predation-risk zone close to shore. By the fall of that year, I had acquired the RAP system and was ready to test its operation on white sharks at the island.

On October 13, 1997, I was standing on a sandy beach looking out on Año Nuevo, a flat rocky island only a half mile from Monterey Bay, south of San Francisco in Central California. The island is roughly a quarter mile long and 100 yards wide at its southern end and becomes narrower toward the north and curves offshore to the west. Barely visible on the island in the mottled gray and white fog were several tall, old wooden buildings, the remnants of an abandoned Coast Guard station. A slight breeze from the direction of the island brought with it the strong, pungent smell of guano, deposited over the years by the mass of seals, sea lions, and seabirds that make the island their home.

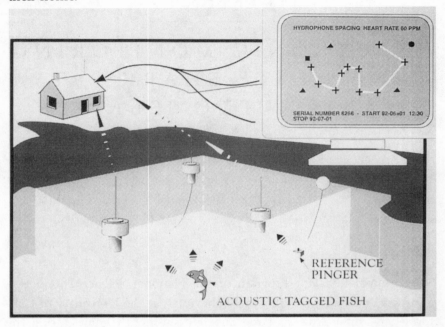

Shown here is a diagram of the radio acoustic positioning (RAP) system, which was set up in the waters near Año Nuevo Island. A transmitter, attached to a shark, sends out an ultrasonic signal that radiates outward underwater and is detected by a hydrophone at the bottom each of three buoys moored in a triangular array. The electronic tag is shown in the diagram on the fish with its pulsing signal indicated by arrows radiating outward toward the buoys. Upon detection of a particular ultrasonic pulse, each buoy emits a radio signal into air from its antenna that is received by another antenna mounted on the roof of a building on the island. The signals, received at this base station, are processed by a timing module interfaced to a computer that instantaneously displays the positions of the shark relative to those of the buoy before saving the data on a hard disk. Shown in the upper right-hand corner is a diagram of the computer monitor with crosses denoting each position of the shark—these are attached by lines to form a track. The buoys are evident on the monitor as triangles. Measurements recorded by sensors on the tag (swimming depth and stomach temperature were recorded by us) are shown on the monitor—a rate of heartbeat is included in the diagram. We simultaneously kept track of five sharks, recording their rate of swimming and their degree of separation, for an average of ten hours per day. This was no mean feat. Think of the effort required to record the movements of five of your work associates and their interactions during an eight-hour work day. (REDRAWN FROM PRODUCT BROCHURE, VEMCO LTD, HALIFAX)

Large waves slowly rolled toward us, their rounded peaks rising as they approached and then broke, forcing water to rush up onto the beach through our feet. A large elephant seal, pushed toward the beach by the onrushing water, looked at us curiously with his huge eyes, almost the size of golf balls, which enabled it to see well in the dimly lighted depths of the sea. Standing next to me was Pete Dal Ferro, an undergraduate student employed by the Long Marine Laboratory of the University of California at Santa Cruz. His job today was to take me to the island in an inflatable raft, where we would test the operation of a sophisticated radioacoustic positioning system installed to monitor the hunting behavior of white sharks around the island.

We now dragged an inflatable raft out from the space between the two decks on a double-decked truck and carried it to edge of the water, where we left it with its pointed bow facing into the waves. Pete lifted the heavy outboard engine off the truck, carried it over to the stern of the inflatable, and attached it with its shaft and propeller elevated above the sand. I climbed to the top deck, picked up a huge lead-acid battery, twice the size of an ordinary car battery, and handed it down to him. He walked slowly toward the inflatable, hindered by the weight of the battery, and put it inside the inflatable. We were both happy that the second battery was already on the island. These two batteries would power our equipment for close to a week, and could be charged using a diesel-fuel generator when we visited the island two to three times a week. I next handed down a waterproof plastic bag holding my clothing, and this was placed on the floor of the inflatable. We were accompanied by Callaghan Fritz-Cope, an aspiring nature documentary maker and member of the Pelagic Shark Research Foundation, an organization promoting shark research in the Santa Cruz area. The three of us wore wet suits because the drive to the island was going to be wet this morning as we negotiated the large breaking waves.

Pete, Callaghan, and I now moved to either side of the inflatable, grabbed its rubber handles, and waited for a moment between waves in order to drag the rubber raft out to sea. A large wave broke and water rushed up the beach, lifting the inflatable off the ground. We charged forward, holding the inflatable toward the next wave. Pete adeptly jumped into the raft, dropped the engine, and, with a quick pull of a cable, started it. Callaghan and I then jumped aboard, and we sped off through the surf zone, quickly passing over the next wave, which was growing larger and ready to break. As hard as we tried, we couldn't avoid being lifted into the air and coming down on the water with a back-jolting crash. Water splashed everywhere. I thought to myself, "This is not good for one's back," and rose slightly on the pontoon so that my legs could absorb the next crash. Once out of the surf zone, the waves were smaller and the inflatable, after passing over each wave did not come down on the water with such force. With experience gained from countless trips, Pete drove the inflatable toward the southeast behind the island. Here the seas were calmer because there was protection from the huge northwesterly swell that crashed into the seaward coast of the island. Sea lions dashed off to the right and left in order to avoid being hit by us as we crossed the channel. The cacophony of sea lion barks and seal grunts grew louder as we approached the landward side of the island. The water near the beach was crowded with young elephant seals, their heads bobbing up and down in the water as they stared at us. We drove the inflatable up onto the beach on the crest of a wave, quickly removed the gear, carried the raft up the beach, and deposited it where it would not be carried away by the high tide that would occur later in the day.

With our gear in hand, we now climbed up a steep but low cliff at the edge of the beach, and slowly trekked across the terrace of the island on a wooden walkway toward a tall wooden building on the other side. On either side of us was a seemingly endless mass of sea lions and seagulls. We opened the door and walked into a cavernous building with a ceiling two stories high. I removed my wet suit, put on dry clothes, and washed my hands with fresh water kept in a plastic flask—one wanted to avoid rubbing the salt from one's hands on electronic equipment. Salt attracted water, and the two together would conduct electricity and short an electrical circuit. I walked across the huge room and over to a steep wooden ladder with metal railings next to the wall on the side of the building facing the ocean and began climbing up to the loft where our equipment was stored. Reaching the top of the ladder, I opened the door and walked into a small room with a bed on one side and a large picture window looking out on the waters surrounding Año Nuevo. Neatly arranged on a shelf in front of the window were a laptop computer and several other instruments comprising the radioacoustic positioning system. Looking outside the window, I could just make out three buoys that slowly moved in and out of sight as they rose on the rounded crest of a wave and fell into the trough as the massive waves slowly made their way toward shore.

Earlier that week, several of us had carried three large buoys in the research vessel of the Long Marine Laboratory to Año Nuevo and moored them in a triangular array in waters 60 to 80 feet deep in front of the island. Two buoys, one a third of a mile off the island's northern end and a second the same distance from the southern end, formed the base of the triangle. The third buoy, the apex of the triangle, was placed midway between the other two buoys and a quarter mile farther offshore to the west.

These were not ordinary fishing buoys, but were sonobuoys, buoys containing electronic instrumentation for recording underwater sound. Each buoy consisted of a fiberglass cylinder, a foot and a half across, that housed an ultrasonic receiver, a signal processor, and a radio transceiver (the prefix "trans-" indicating that it received and transmitted signals). An electronic cable led from the receiver to a plastic pipe mounted on the bottom of the buoy with a hydrophone at its end. The transceiver was connected to an antenna mounted on the top of the buoy which was 3 feet high. There was a thick ring of foam flotation around the center of the fiberglass cylinder. The hydrophone and receiver detected the *ping-ping-ping* from a transmitter on a shark; the signal processor possessed a clock (synchronized among buoys) that timed the arrival of the pulses, and the transceiver and antenna sent radio packet with this information to a receiving station situated onshore. This equipment had been installed at this "base" station on the previous day. It consisted of an antenna high on the roof above the window, a metal box with another transceiver, and a laptop computer. The computer would plot the position of the shark relative to the buoys on its screen and display measurements of the shark's swimming depth, speed, or stomach temperature, provided the transmitter carried the appropriate sensors.

I lifted my binoculars to my eyes and gazed southward along the coast toward Santa Cruz. I could barely make out a small cabin cruiser in the distance, moving up and down as it pounded its way through the waves on its way toward the island. Driving the boat was Sean Van Sommeran, who had started the Pelagic Shark Research Foundation (PSRF) to promote research and conservation of sharks. He was a folk hero among the surfing and fishing communities of Santa Cruz. Sean was fascinated by sharks and was eager to help me with this research project. Standing in the boat next to him was Scott Davis, a graduate student at the Moss Landing Marine Laboratories in Monterey, who had come to Northern California with one mission—to study white sharks at Año Nuevo. Along with Callaghan Fritz-Cope, they had worked with me day and night during the last week setting up the radioacoustic positioning system. They were now going to help me determine the range over which the RAP system could detect a shark carrying one of our ultrasonic transmitters.

Upon arriving in front of the island, Sean called by radiophone and asked for further instructions about his role in the range tests. I told him to drive the boat to the center of the array where he would be equally distant from each of the buoys. He should then lower the line, with a transmitter attached just above a weight, halfway from the bottom, and slowly begin to motor offshore. This would simulate the cruising movements of a white shark. I would then track the movements of this imaginary shark with the RAP system as it swam away from the array so that I could determine the farthest distance at which the transmitter (and shark) could be detected.

As they made their way toward the center of the buoy array, I turned on the laptop and placed my hand on the computer mouse to click on the icon that opened the program operating the RAP system. I then instructed the program to find the position of the buoys. The screen would display a vertical dashed line for the y-coordinate and a horizontal line for the x-coordinate—where the two lines crossed was the center of the triangular buoy array. This would serve as a reference for the positions of the sharks. The transceiver sent out a signal that synchronized the clocks of the three buoys and instructed the buoys to determine their relative positions. A low-frequency beacon on sonobuoy A transmitted a pulse and the other two sonobuoys, B and C, listened for the pulse. Then sonobuoy B transmitted and sonobuoys A and C listened for the pulse, and finally C transmitted its pulse with sonobuoys A and B receiving the pulse. A light-emitting diode (or LED) on the transceiver lighted up each time instructions were sent to each buoy. Suddenly, three small black triangles appeared on the screen for the three sonobuoys. In a corner of the screen, the distances between the sonobuoys were displayed—the distance between A and B, the two sonobuoys at the base of the triangle and nearest to the island, was 600 meters, almost a third of a nautical mile. (This type of mile, based on the circumference of the earth, is equivalent to 1,852 meters, and is 1.14 times the length of the landlubber's statute mile.)

Sean told me that the beacon was now in the water. I moved the cursor to the top of the screen and pressed down on the mouse, and a series of options was displayed, one being RECORD DATA. I selected this option and waited with eager anticipation for the position of the beacon to be displayed on the monitor in the center of the triangular array. Suddenly, to my great relief, a symbol appeared in the center of the array indicating the position of the imaginary shark. One by one, symbols on the screen appeared as the imaginary shark slowly swam away from shore. I noticed that the track of the imaginary shark on the screen appeared to pass between sonobuoy C, which was the buoy farthest west and formed the top of the triangle of buoys, and sonobuoy B, which was the most northern buoy and was on the right-hand base of the triangle. The simulated shark was shown to travel here because the pulse from the beacon took longer to reach the farthest buoy, A, than the next-farthest buoy, B, the two buoys at the base of the triangle.

As the boat reached a distance of 1,200 meters (three-quarters of a mile from shore), the outer boundary of the zone of high predation risk, the base station no longer plotted the position of the imaginary shark. I pressed a key to select a page displaying current information about the sonobuoys. Continuously appearing on the screen were the numbers of pulses from the beacon detected by each of the three sonobuoys during each listening period of a dozen seconds. Now only sonobuoys C and B, which were nearest to the beacon, were detecting pulses. The beacon must be within the range of all three sonobuoys for its precise position to be plotted by the RAP system. However, I could at this time still determine that the imaginary shark was present near the island—all it took was for the beacon to be detected once by any one of the three buoys.

My attention, now directed at the computer, was suddenly drawn to Sean's voice on the radiophone: "Ahoy, Pete, there's a shark attack right in front of you. Look out of the window to the right and you'll see it. It's not more than fifty yards from your window." I looked out the window and, sure enough, there was a large pool of red blood less than 100 meters from shore. Swimming slowly in the bloodstained water was a large white shark—much of its back along with its dorsal and tail fins were out of the water. Scott Davis was now slowly driving the boat toward the shark. Sean asked, "Can we tag the white shark?" I noticed Scott had now stopped near the shark and dropped a seal decoy into the water. This was a board of plywood, cut in the shape of a seal, which was used to lure a shark close to the boat so the shark could be tagged with a pole spear. I replied, "We know that the RAP system will detect the beacon from the range test just completed. Why don't you go ahead and tag the shark." Sean replied, "Roger, will do!" His replies during tense situations often assumed a military tone, which I enjoyed. I shared his fascination with military history.

The decoy was now drifting closer to the pool of bloodstained water, but the shark was no longer visible on the surface. A minute later, the head of the shark slowly rose out of the water a couple of body lengths away from the decoy, and it began to swim toward the decoy with sweeping beats of its tail. Scott slowly pulled in the line attached to the decoy until it was at the side of the boat, and then he pulled the decoy toward the bow, inducing the shark to swim alongside the boat in an ideal position to for Sean to tag it. The shark was large, roughly 17 feet long, more than two-thirds the length of the 22-foot boat. Sean now reached over the side of the boat with the pole with the tag at its end in his hands, and his arm jerked downward as he appeared to apply the transmitter to the shark. Seconds later, he said over the phone, "Transmitter applied!" It was now 3:40 P.M. in the afternoon. I replied, "Keep an eye on the shark when it next comes to the surface. Look at the application wound to see whether the barb has stayed within the shark's body." The shark soon surfaced again, and Sean informed me that the transmitter was visible, attached to its back.

I now turned toward the laptop computer and selected the option RECORD DATA. I waited to see if the RAP system would record the position of the shark. The next half an hour passed slowly because not a single position of the shark was displayed on the monitor. Could the shark be outside the range of the monitor? The shark was tagged south of the island some distance from the sonobuoys. Then a small solid circle appeared on the screen between the southernmost buoy and the island, then another, and so forth to form the track of the shark slowly swimming along the coast of the island, waiting to intercept the next seal trying to reach the island. I shouted, "Hallelujah!" and immediately informed Sean and Scott that the system was recording positions of the tagged shark.

The boat with Sean and Scot had now drifted farther south of Año Nuevo Island, and the cliffs of the island blocked it from my view. Sean replied that they were elated that the system was working, because there were two more sharks swimming around the boat that could be tagged. I told them to go ahead and tag them, and thought to myself that this was just too good to be true. It usually required a Herculean effort to tag the subjects of one's study. Scott lured the second shark, also 17 feet long, to the boat with the decoy, and Sean tagged her at 4:30 P.M., less than an hour after tagging the first shark. Scott filmed the shark rolling on its side as it passed the boat. We later looked at his video and noticed the absence of a clasper along the margin of its anal fin—the shark was a female. Half an hour later at 5:00 P.M., Sean tagged the third shark, which was smaller than the other two, roughly 15 feet long. We waited another half an hour, in case a fourth shark might rise to the surface and be tagged. That, I suppose, was asking for too much on a single day on which we had not intended to tag sharks at all. By now the winds from the northwest had become stronger. Huge rounded waves were rolling toward shore. The buoys slowly rose on the crest of each wave and disappeared in the trough of the same wave. By 5:30 P.M., the seas had grown so rough that Sean and Scott left, slowly proceeding on their hour-and-a-half journey back to the Santa Cruz marina.

I would have to drive ashore in the inflatable, pack the inflatable in the truck, and return to the Long Marine Laboratory. Before doing this, the RAP system had to be instructed to monitor the three white sharks continuously in our absence. I selected several options in the program so that positions of the first shark would be recorded for a dozen seconds, then positions of the second shark for a similar interval, and then positions of the third shark, and so forth. The system would reestablish contact with the same shark every thirty-six seconds and record its positions then for a period of twelve seconds. I then checked all of the battery connections to make sure that they were secure. Everything appeared in good shape. Pete Dal Ferro had come up to the loft upon hearing that shark attack had occurred and helped me communicate with Sean. Callaghan had recoreded all of my actions, and those of Sean and Scott in the boat with his video camera. We now climbed down the ladder, put on our wet suits, and started walking toward the beach where we kept the inflatable. It had been a long and hard day, but an extremely gratifying one.

We succeeded in tagging two more white sharks that fall—a 14-foot-8-inch female on October 16 and a 15-foot-5-inch female on October 26. Both sharks were tagged off the northern end of the island, in contrast to the three earlier sharks, which were off the southern end of the island. Every three days during the next two weeks, one of us visited the island to save the data collected by the RAP system on the laptop's hard disk and back up the file on a disk drive. We also turned on the diesel generator in the building and recharged the two heavy-duty marine batteries so that they would operate the system for the next three days. Rough weather precluded further operation of the RAP system after October 30.

Our overall objective was to describe the movements, behavior, and interactions among sharks while they searched for seals near Año Nuevo Island. At the Farallon Islands, we were able see the sharks only when they were above the surface, and from these observations inferred that the sharks captured their prey using a particular tactic. The RAP system would record information about the behavior of the sharks while they were beneath the water. We hoped to use this system to determine how social sharks were when hunting for seals. During the five-year study at the Farallons, we observed a single shark on 144 of the 195 days that white sharks were observed either feeding on pinnipeds or investigating a surfboard, which was allowed to drift away from the island and function as a lure to attract sharks. However, we also observed two sharks on 43 of the days and three sharks on 3 days. Might these latter sharks be hunting for pinnipeds in a coordinated way? On one occasion, two sharks were observed to chase a sea lion in front of the island. However, I was uncertain whether the two sharks were cooperatively chasing the prey or whether the second shark simply became aware of the attack only after the first shark begun to chase the prey. When more than one shark was observed at a kill at the Farallon Islands, it was usually to feed on a recently killed seal or sea lion, which was floating immobile on the surface. At these times, the sharks splashed water at each other with their tails in what I concluded was a threat display, as discussed earlier.

Simultaneous tracking of several animals at one time is certainly a formidable task. Imagine continually monitoring five employees during an eight-hour work period (a third of the day), recording the rates at which they walk around the building, how far apart they are from each other, and what they are doing when they come together. The five sharks tagged and tracked during the fall of 1997 were detected in a half-mile-square area an average of 9.6 hours per day during a two-week interval by one of the three sonobuoys.

Kelly Cantara, a research intern from Southampton College in New York, who devoted six months in my laboratory to analyzing the records collected by the RAP, learned much about the five sharks. To start with, they patrolled the waters around the pinniped colony at all times of the day. Let us imagine that the sharks were to visit the island an equal number of times every hour of the day—one–twenty-fourth of their day would be spent at the island during each hour. There was no difference between the time spent at the island each hour and this imaginary fraction. The white sharks visited the island as often during nighttime as during daytime.

Why be at the island at night, if not to feed on seals and sea lions?

We never observed nocturnal feeding by white sharks at Southeast Farallon. However, the reason for this may have been the absence of the shark watch at night, when we couldn't see the waters surrounding the island at night. There was little light to illuminate a shark if it were to feed on a seal or sea lion at the surface of the water. Most evidence to date had suggested that the white shark is solely a daytime predator. We observed white sharks feeding on pinnipeds only during daytime at the Farallon Islands simply because the sharks could be seen from Lighthouse Hill only during the day. We couldn't eliminate the possibility that the same sharks might be feeding at nighttime.

The five sharks tracked at Año Nuevo Island always patrolled close to shore. The sharks could only be tracked when in front of the island within the range of all three buoys. Imagine a large circle drawn around each buoy, indicating its range of signal reception. The shark could be detected only where the three circles overlapped. It is a common practice to describe an animal's home range by connecting its most distant positions to form the smallest, or "concave," polygon. We made this type of polygon for the five white sharks—this area encompassed by the polygon extended from the shoreline to 700 meters (766 yards) from shore despite the RAP system's ability to track a shark at much greater distance from the coast of the island. Twenty-four-hour plots of the movements of the sharks indicated that they generally swam back and forth along the shore. They patrolled the shore during day and night, approaching within two meters of the shore—the limit of the system's positional accuracy. The sharks spent most of their time where they were ideally situated to intercept seals and sea lions departing and returning to the pinniped rookery on the island.

There was no evidence that any of the five sharks had exclusive territories. If the sharks favored separate areas, they might have been detected by one of the sonobuoys more often than the other two sonobuoys. We compared the numbers of fifteen-minute periods in which the sharks were detected at each of the three buoys to an equal distribution of detections among the buoys, and found no significant difference between the two. John Richert, another research intern from Southampton College in New York, produced a contour map of the frequency that each shark was in 100-meter, (110-yard)-square areas in front of Año Nuevo Island. The peak in the activity of the shark that was first tagged was east of the center of the island. The sharks tagged second and third had overlapping peaks of activity, one off the northern and another off the southern end of the island. Although each shark favored a slightly different area than the others, all five sharks frequently moved over the same area.

Were the sharks social or solitary hunters? The sharks moved in and out of the range of the array simultaneously at some times, consistent with their moving in a pack, but at other times arrived and departed separately. One might ask whether the sharks tagged at the same kill accompanied each other more often than sharks tagged on different days. These sharks might be members of a hunting group. However, our tracks of the sharks did not support this conclusion. We did not find that the three sharks, tagged while collectively feeding on a sea lion, were detected more often together than with the fourth shark, tagged three days after the predatory attack, or with the fifth shark, tagged thirteen days later. You might expect the first three sharks to spend more time together if they related to each other and hunted in a coordinated manner in a pack.

There was little consistent evidence of sociality among the sharks when they were near the island. It was very rare that the sharks approached each other. The median distances that separated each of the five sharks ranged from 80 to 420 meters (88 to 460 yards). If the tag mates formed a social group, you would expect shorter separation distances among the first three sharks, who were tagged on the same day, than between the two sharks tagged later during the study. This was not so. For example, the separation distance between the first and fifth tagged sharks was less than among the first and second and third sharks.

If sharks were attracted to each other, the true distances between individuals should be less than the random separation distances. If avoiding each other, the true distances should be greater than the random separation distances. Much of the time, there was little difference between true and random separation distances; sometimes the true distances were less than the random ones, and at other times, the true distances were greater than the random distances. Thus, the sharks were neither attracted to each other nor did they avoid each other while swimming near the seal colony. They were most likely searching for prey by themselves, yet the others might be attracted to the site of the kill once one of them captured a seal or sea lion.

Only two predatory attacks, one during nighttime and another during daytime, were detected during the fifteen-day period that the five sharks were monitored during 1997. On October 16, the shark that was first tagged exhibited two bursts of swimming consistent with chasing prey. These swimming accelerations were apparent in the close spacing of three or four points in the track of the shark during twelve-second intervals at 11:06 and 11:07 P.M. The shark then turned and swam slowly in a straight line until 11:26 P.M., its movement possibly hindered by the added burden of carrying the prey in its mouth. A tight cluster of positions near a sonobuoy indicates that the shark then fed on the pinniped from 11:16 to 11:24 P.M. The shark then swam away from its prey in a clockwise circular path to a distance of 600 meters (a third of a mile) before returning briefly at 11:58 P.M. At this time, the shark appeared to feed a second time, based on the cluster of positions there. Another white shark appeared at 11:49 P.M., 200 meters away from the first shark, approached the shark nine minutes later, and then left in a westerly direction. I suspect that the two sharks exchanged aggressive displays at this time, with the shark that initially killed the prey being the loser and, for this reason, first leaving the area. It is difficult to envisage what happens during an attack based on these inferences, or indirect conclusions. We needed to see the smoking gun, or more direct evidence of feeding.

A more reliable indicator of feeding would be a sudden elevation of the stomach temperature, indicating that the shark had swallowed a warm-bodied seal. Since the body temperature of white sharks varies between 73° and 79°F and the temperature of the elephant seal is a constant 100°F, the stomach temperature of the shark might be expected to warm as much as 20°F after eating a seal. The duration of stomach warming might also be an indication of the size of the meal. A juvenile elephant seal, weighing 200 pounds, would take less time to cool down than an adult weighing half a ton.

We decided to introduce a transmitter into the shark's stomach with temperature and depth sensors to get a more reliable electronic "fingerprint" of its feeding behavior. Again, we anticipated seeing on the computer monitor a short, straight trajectory composed of closely spaced positions, indicating the shark swam rapidly toward the seal, that to be followed by tightly clustered positions indicating that the shark stayed in one place feeding. Now, however, displayed on the monitor alongside of the track of the shark would be alternating measurements of the shark's depth and stomach temperature. If the shark ambushed its prey, there should be a rapid decrease in swimming depth as the shark dashed to the surface to seize the seal. The shark's stomach temperature would later rise after swallowing the warm parts of the seal.

We reasoned that it would be best to monitor one shark at a time with the RAP system if we wanted all of this information. Previously, we had recorded the positions of a single shark for over a dozen seconds before moving on to the next shark. A full minute would go by before the RAP system passed through the five sharks and began listening to the same shark again. A lot could occur during an interval of a minute. That shark might even dash to the surface and seize the seal in a minute's time and we would not pick up that activity.

We set up the RAP array at Año Nuevo Island again during the second week of October 1998. We began making daily trips to the island in the small research boat of the Long Marine Laboratory, seeking to tag a single white shark. Our daily research activities were being filmed for an episode of *National Geographic Explorer* entitled "Tracking White Sharks at Año Nuevo Island." Burney Le Boeuf, the seal behaviorist at the University of California at Santa Cruz, joined me on these trips. I was filmed talking about the predatory behavior of the white shark, and Burney, the evasive tactics of the northern elephant seal. John Francis, a former Ph.D. student of Burney, was the director of the film.

Once in front of the island, we would drive the boat to the RAP buoy at the northern end of the island, turn off the engine, and let the boat drift in front of the island with a seal decoy tethered to the boat.

The winds, usually coming out of the northwest, pushed the boat across the front of the island toward the southern sonobuoy in the RAP array. This area was in the middle of the shark attack zone. We spent two days drifting back and forth in front of the island without attracting a single shark to the surface.

It was on our third visit to the island, on October 22, 1998, that we encountered our first white shark. It was a foggy day, and we could barely see the window to the loft that was our base station no more than 100 yards away on Año Nuevo Island. We were in a small boat drifting across the island for the second time when suddenly an 18-foot-long white shark rose up out of the water, mouth opened wide, and extended its upper jaw over the decoy and its lower jaw under the decoy. We all stood there awestruck, waiting for the shark's jaws to close, breaking the flimsy decoy in half. The cameraman was also distracted; he did not turn on his video camera right away. The shark hesitated in this dramatic pose for the longest time before surprising us—it gracefully slid back into the water without demolishing the decoy. Burney shouted, "That's not *Jaws*, that's a picky feeder!" That remark provided me with great inner satisfaction—the reaction of another behaviorist had confirmed an observation of mine made half a dozen years earlier.

I then placed a temperature and depth-sensing transmitter that was the size of a small flashlight with C-cell batteries on top of a large flat piece of whale blubber. One of the barbs on the treble hook was passed through the tissue and half of the blubber was folded over on top of the other half in order to keep the transmitter out of the sight of the shark. I then passed a large sailing needle with waxed line back and forth through the upper and lower ends of the piece of blubber to keep it secure. After tying one end of a line to the blubber, I gingerly placed the whale blubber in the water and let it drift away from the boat. The shark rose to the surface not more than five minutes later and swam toward the decoy, which we pulled toward the boat. We could see that the shark was a female because it lacked claspers. She was roughly 18 feet long and had a large triangular notch in the back edge of her large black dorsal fin. Hence, she had been given the name

Top Notch when she was first observed at Año Nuevo. Top Notch saw the piece of whale blubber while swimming slowly behind the decoy. She turned toward it, opened her mouth, and swallowed the bait at 9:01 A.M.

The equipment that processed signals from the sonobuoys was now aboard the small boat. The transceiver and laptop were kept in a large carrying case in the stern of the boat; the antenna that received the radio signals from the buoys was fastened to the mast that held the boat's radar. I bent down over the carrying case and draped a black towel on top of my head and over the edges of the case to keep the sunlight out. I could now see what was displayed on the monitor of the laptop. A small solid circle appeared on the screen next to the north-ernmost triangle, indicating the position of a sonobuoy. The shark was still close to us because we were close to that buoy. There in the upper right of the screen was the depth of the shark—22 meters (72 feet). I asked Sean Van Sommeran, skipper of the boat for that day, to find out how deep it was here using the boat's sonar. He promptly informed me that it was 24 meters (79 feet) deep—the shark was now swimming close to the bottom. I now looked at the latest measurement of stomach temperature, displayed underneath the depth: the shark's stomach was only 60°F. That puzzled me, but only for a moment. The whale blubber was still partly frozen when swallowed by the shark, and it had yet to thaw out. Over the next hour, the temperature slowly rose to 68°F. I now looked at the shark's depth—it had changed to zero. The shark had just performed the stereotyped approach behavior (swimming along the bottom and rising to the surface at a steep angle) and must now be on the surface. "Ahoy, shark on the surface," yelled Sean. This struck me at the moment as a testament to the power of technology—the computer screen had alerted me that the shark had surfaced before Sean had seen it.

We returned to Santa Cruz by noontime in order to transport the RAP system to Año Nuevo Island. We took the equipment to the island in an inflatable and installed it again in the loft overlooking the waters of Año Nuevo by 6 P.M. on that same day. The stomach tem-peratures now recorded hovered around 77°F, well within the normal range of temperatures to be found in the shark's stomach. The frozen whale blubber had by now completely thawed.

We hoped to detect when Top Notch fed by the slow rise in stomach temperature as the warm body of a freshly killed seal warmed her stomach. She was tracked intermittently for twelve days between October 22, and November 18, 1999. She spent an average of five hours each day patrolling in front of the seal colony within the range of the sonobuoy array. There was little evidence of her feeding when the sonobuoys were in place—twice the sonobuoys broke free and had to be replaced. Her stomach temperature remained above 79°F during the longest period of continuous recording: six days, from October 31, to November 5.

How active was she when searching for a meal off a seal colony? Let us examine her hunting behavior during a single day, November 2. She spent thirteen hours of that day within the range of the RAP system. She repeatedly swam back and forth 200 to 300 meters from shore along a north-south axis, where she was ideally positioned to intercept any seal or sea lion departing and returning to the rookery on the island. Top Notch left her "picket line" only once during the day around noon, and she swam away from the island in a southwesterly direction. She moved along the path generally taken by a pinniped moving to and from Año Nuevo along an underwater channel. When she was farthest from the island, she dashed to the surface, swimming at a rate of 4 meters per second, as if trying to ambush a seal, but no elevation in her stomach temperature was detected by the RAP system—she had not succeeded in catching a seal. She again became active between 7:00 P.M. and midnight. She twice accelerated as though chasing a seal, once at 9:02 P.M. and again at 9:43 P.M. She took off with a speed of 6 meters per second (11.7 nautical miles per hour) toward the surface, and there was a temporary decrease in her stomach temperature, possibly due to cool water entering her mouth held open in an attempt to seize the prey. She then dashed downward toward the bottom with a speed of 7 meters per second, but no change in stomach temperature was recorded.

Although all six of the sharks that we monitored during 1997 and 1998 stayed in the high-predation-risk zone much of the time, the

RAP system provided little evidence of their feeding. This suggests that individuals expend considerable effort searching before they succeed in catching a seal or sea lion. Only two of the five sharks monitored during October 1997 appeared to feed—two at night on the sixteenth and one during daytime on the eighteenth. We never recorded a prolonged elevation in Top Notch's stomach temperature indicative of feeding during the twelve days she was monitored during fall 1998. It appears that white sharks expend considerable time and energy patrolling the shores of a pinniped colony in order to catch a single seal or seal lion.

If it is so difficult to capture prey, one obvious feeding strategy would be for white sharks to remain close to each other so that, if one shark caught prey, the others could scavenge upon the leftovers. A 15-foot white shark can theoretically survive a month and a half on a single 66-pound meal of whale blubber. A one-year-old seal, weighing roughly 300 pounds, with half its weight (150 pounds) in its exterior layer of fat, would have over thrice the energy necessary to sustain a white shark for a month and a half. Minus the 66 pounds consumed by the shark that initially captured the seal, the carcass would contain 89 more pounds of fat, providing sufficient energy to sustain a second shark for over a month and a half, while the 150 pounds of muscle could sustain a shark for a similar period. Of course, each shark would ingest fat with muscle when eating the seal. A single juvenile seal could thus provide two of the shark's companions with sufficient energy to survive the same period without feeding. It is therefore not surprising that the five sharks tracked during this study did not stray that far away from each other. By staying close to each other, they can compete for a share of the rest of a carcass after another shark seizes the seal and removes the first bite. This might explain why a noninjurious ritual combat evolved, in which individuals slap water at each other in order to determine who feeds on the rest of the kill.

What more is there to learn about the white shark? To start with, we need to know where they go when they are not at a seal colony. There is a another type of electronic tag, called a data-storage, or archival, tag, mentioned previously, which is a microprocessor-based instrument with sensors that record measurements of various environmental properties (light, temperature, and depth) and which stores these measurements in an electronic memory until later removal from the animal. These tags infer the daily geographic positions of a tagged fish from these physical measurements, an ability termed "geolocation." The first generation of geolocating archival tags had to be recovered when the tagged fish was later captured. This type of tag could not be used with the white shark because there was neither a commercial fishery nor a sports fishery for the species off the western coast of North America. In short, there was no way of recovering the tags from captured sharks. The California legislature had provided the species with temporary protection from capture in 1993 with Assembly Bill 522 and five years later had legislated continued protection. However, the latest generation of "pop-up" archival tags release from the fish after a set time interval, float to the surface, and transmit a record of prior positions (with some depth and water-temperature information) of the fish to a satellite. This, in turn, transmits the information to a receiving station situated on land. A graduate student at Stanford University, Andre Boustany, and five colleagues placed six of these expensive tags on white sharks at Año Nuevo and Southeast Farallon Island. The tags on two sharks indicated that they remained close to shore in shallow water the entire year. The four other tagged sharks moved offshore, where they remained in deep water for four to six months. One of these sharks traveled over 2,000 miles to the Hawaiian Islands.

Another pressing question is whether white shark predation controls the growth of seal and sealion populations along the coast of North America. It may be that this single species has slowed the once rapid growth of pinniped populations along the western coast of North America and, in doing so, has prevented pinnipeds from limiting the populations of fishes that we depend on for food. This type of ecological control is termed "top-down" because a species higher on the food chain limits the abundance of another species lower on the chain. This effect is in opposition to "bottom-up" control, in which species lower on the food chain controls the abundance of ones higher up on the food chain. For example, strong winds may cause deep, nutrient-rich water to rise to the surface, enabling microscopic plants to proliferate, and production on this trophic level permits the next higher link in the chain, microscopic animals, to grow in numbers, and so forth.

The importance of bottom-up regulation in the oceans has been well established; however, the prevalence of top-down regulation is less well known. The sea otter is the main player in the best-known example of top-down regulation. Sea otters had been hunted to near extinction by the middle of the twentieth century, and for this reason they were one of the first species protected by the Marine Mammal Protection Act. Once sea otters were no longer killed, their numbers grew and, at the same time, so did the size of the kelp forests along the western coast of North America. The otter exerts its top-down effect on kelp (two links below in the food chain) by eating sea urchins and abalone (one link below), two species that graze upon kelp, and act jointly to reduce the size and extent of kelp forests.

Could white sharks have the same type of effect on the populations of fishes? The first step in answering this question would be to demonstrate that the species limits the abundance of seals and sea lions. We can record the rate of white shark predation on seals at pinniped colonies based on observations of feeding from the surrounding cliffs and the detection of feeding with the RAP array using an indirect "fingerprint" of predation. This would consist of a unique pattern of movement accompanied by a dash to the surface and slow increase in the shark's stomach temperature. The effect of predation, of course, must be compared to the other causes of mortality of seals at the colony. It is possible that seals are captured by white sharks only during the six or seven minutes that it takes them to swim the last quarter of a mile to their colony after their first trip to sea. For the unfortunate ones, it is a fatal voyage. But this is an ecosystem regulating itself. Understanding how such regulation works at every level in the natural world is a key to our survival—particularly if we begin to understand how much damage we can do by unwise interference with nature's self-regulation.

CHAPTER 14

SHARK-EATING HUMANS OR MAN-EATING SHARKS?

Sallie and I, sitting on the center seat of the panga, were busily putting on our scuba tanks as the helmsman drove us just south of the Espíritu Santo Seamount. Sallie Beavers, smart, energetic, and strong-willed, had been an ideal research partner during the last three years. We enjoyed each other's company and worked very hard together. It was the morning of September 12, 1998, toward the end of a cruise aboard a research vessel of the Scripps Institution of Oceanography. The purpose of the cruise was to establish whether the warming of local waters by an El Niño affected the residency of the migratory fishes at Espíritu Santo. We had spent most of our time catching adult 150-pound yellowfin tunas by rod and reel, putting long-term ultrasonic beacons on them, and releasing them at the seamount. We had placed two moorings, one at either side of the seamount ridge. These were equipped with electronic monitors that would detect when the yellowfin swam within their ranges over the next two years.

There was now a little spare time to make a scuba dive in order to search one more time for schools of scalloped hammerhead sharks. We had drifted over the seamount wearing a mask and snorkle each morning of the week, searching for the schools, but had observed only one or two hammerheads at a time. Our bodies fell backward into the water, we placed our regulators in our mouths, and began to swim downward toward a cluster of white fishing buoys visible in the crystal-clear water no more than 10 meters below us. The buoys were at the top of the mooring, which held our monitoring device. Halfway down the line were three more of the small buoys, and a meter below was the electronic monitor, the size and shape of a small fire extinguisher. The purpose of the extra buoys was to hold the monitor up in the water column, where the top cluster of buoys cut from the line.

The massive boulders on the slope of the seamount came into view as we reached the buoys at the top of the mooring. There was a small school of less than a dozen little silver fish swimming around the buoys. We continued downward, looking on our dive computers, until they displayed a depth of 120 feet. We then swam horizontally in the direction that the rocky bottom sloped upward most steeply, leading to the southern pinnacle. We were no more than 10 meters above the bottom. I looked at my dive compass to ensure that we were swimming toward the northwest, the general direction of the seamount ridge.

I glanced to the side to make sure that Sallie was next to me, and then looked forward into the distance. There were no hammerhead sharks here, but this was to be expected. The schools had never been seen over the southwestern slope despite our many attempts to find them here. The northwestern slope of the seamount, 120 feet from the surface, was particularly attractive to the hammerhead sharks. (As explained in Chapter 7, this may well be due to the sharks orienting to a unique property of the seamount's magnetic field there.) A school might swim southward toward the peak of the seamount, but would invariably turn around, and head back to the same place. This was a base of operations, where they remained during the day and returned to repeatedly after making their nightly excursions out into the surrounding environment to feed. Swimming at the surface, I had often observed a hammerhead school performing this same maneuver out of sight of the bottom, and wondered what led the sharks to the area and how they recognized it. They always returned to the same place despite being carried away in the strong currents. The sharks must somehow recognize the local landscape, but what could they detect so far from the bottom. Could it be a unique pattern in the electromagnetic field? I had a video sent to me by a colleague that showed small hammerhead sharks apparently following the lines of current (or electron flow) between two electrodes. These were powered by a small battery. The flow of current simulated the tiny electromagnetic dipole emitted by the prey of the sharks. Could the sharks be responding to a large-scale geological version of the small biological dipole?

But alas, was it not folly to wonder about how the hammerheads navigated around the seamount when there was a more pressing issue, and that was that there were few, if any, hammerhead sharks left at the Espíritu Santo Seamount anymore. I had not seen a single hammerhead when I visited the seamount during the previous summer to make a documentary about the hammerheads. During this cruise, we had set up monitors for detecting hammerheads and other oceanic fishes at many seamounts and islands in the Gulf of California. We had looked for hammerheads on which to place electronic flags at Las Animas Rock, Espíritu Santo Seamount, Las Arenitas off Cerralvo Island, and Gorda Seamount, and had found none.

We were by now drifting over the small rounded pinnacle, less than 10 meters across, which was the highest point on the seamount. This peak was located at the southern end of the ridge. There were only a few loose schools of small red creolefish swimming above the seamount. Sallie pointed at a rocky crevice, from which a huge grouper stared up at us. The bottom now fell out of sight, visible only as a mottled gray and white, before it rose again into sight, revealing the broader northern peak of the seamount, which was roughly 25 meters long. As we drifted over the northern edge of the peak, we watched female creolefish and silver-colored jacks dash upward while pursued by groups of males that released clouds of white sperm into the water.

We now swam downward over the northwestern slope of the seamount. As we descended, it grew darker—the bottom was now a darker gray and the fish that swam near it were not red, but greenish blue. In every direction were large schools of jacks and pompano. We looked at our depth readout on our miniature dive computers and saw that we were 120 feet deep. There, below us near the bottom, was a tight school composed of thirty to fifty large snappers that were each close to a meter long. This species was usually associated with the hammerheads and thought to be a member of the same migratory assemblage of fishes. I now raised my head and looked forward, expecting to see the sharks. There immediately in front of us was a tiny school of hammerhead sharks. I quickly counted them, "One, two, three . . ." There were only eight hammerheads in the school. I tapped on Sallie's shoulder and pointed toward the hammerheads. We both watched as they hurried away as though they were fugitives hiding from something. By now, our air supply was low. We slowly swam upward and stopped close to the surface, stayed there for five minutes to decompress (flush nitrogen out of our tissues), and then surfaced.

I looked at Sallie with a sad face and lamented, "That is all that is left of the once magnificent schools of hammerheads!" She was similarly downbeat: "And I wanted to see at least a hundred hammerheads in the school!" She was well aware that Don Nelson and I had attached color-coded spaghetti tags to sharks in the vast schools a dozen years ago and had estimated that the population at the Espíritu Santo Seamount exceeded five hundred hammerhead sharks. I replied, "We saw a testament to effectiveness of the long lines and gill nets in the fishermen's pangas." Salvador Jorgensen, now a graduate student at the University of California at Davis, began visiting the seamount four times per year while an intern to perform a census of relative abundance of its various inhabitants. He makes both free and scuba dives at the site during each visit. He has seen few hammerheads at the site during the last three years. In fact, it is imperative that yearly records be kept of the number of fish of various species caught as well as degree of fishing effort in the gulf by both artesianal and sport fishermen. John Richert, a graduate student in my laboratory, is now contacting local fishermen (in one case the proprietor of a smoke-house) and asking them to keep records of their daily catch and effort. These records, once collected on a regular basis, will be entered into a database so that the abundance of species can be monitored from year to year in the gulf. The collection of this type of information is the first step in the effective management of fisheries—it enables you to decrease fishing pressure if a particular population (or stock, to fishery scientists) declines in abundance, as have the hammerheads.

The plight of the hammerhead shark is not unique. People like to eat sharks, species whose life-history characteristics make them sensitive to overfishing and whose wide travel makes fishery management difficult. This appetite has resulted in the recent growth worldwide of fisheries for shark species. These fisheries have a boom-and-bust history. Shark populations have traditionally decreased sharply after periods of intense fishing pressure.

Examples are the fisheries for the porbeagle shark in the North Atlantic, the soupfin shark in the California Current, the basking shark in the waters off Europe and Canada, and the spiny dogfish in the North Sea and off British Columbia. For example, in 1961, Norwegian fishermen began fishing commercially for the porbeagle, a relative of the mako and white sharks. The yearly landings of this fish- and squid-eating shark that lives in the northeastern Atlantic rose to 8,060 tons in 1964, only to fall during the next three years to 207 tons. The yearly landings have not exceeded 100 tons since the late 1970s.

Basking shark fisheries have collapsed in both the Atlantic and Pacific Oceans. A localized fishery for the species began in 1947 off western Ireland near Achill Island. Between 900 and 1,800 of these enormous plankton-eating sharks were captured each year from 1950 to 1956. The catch decreased to 119 per year from 1959 to 1968. The fate of the soupfin, a close relative of the reef shark, is similar. The need for high-grade oil by the military during World War II created a market for the species. The liver, impregnated with oil, can reach a third of the body mass of adults. The price of soupfin oil rose from $50 per ton in 1937 to $2,000 per ton in 1941. The catch of this species grew from 270 tons per year in the early 1930s to a peak of 2,172 tons in 1941 and then dropped to 287 tons in 1944.

Commercial fisheries for sharks have operated in the United States since the 1930s, but they were originally small-scale and restricted in area. They did not begin to grow until the late 1980s. Massachusetts has the largest fishery for sharks in the United States. This fishery mainly targets a single species, the spiny dogfish, which lives in large schools off New England in the spring and summer, and then migrates to the waters of the southeastern Atlantic during winter. It has been served as "fish and chips" by restaurants in the United

States and Europe. The landings of this small but very abundant species were still rising by the late 1990s.

Excluding dogfish from the catch, Florida has the largest shark fishery. This fishery targets many coastal species in the Gulf of Mexico. The yearly landings of these reef sharks rose until the early 1990s, when they began to decrease. The growth of the commercial shark fisheries during the late 1980s in the Atlantic Ocean and the Gulf of Mexico was partly due to a new public appreciation of the value of these species as food. However, a more important reason for the expansion of the fishery was the demand for fins to be used in shark fin soup, a delicacy in Asia.

There are several reasons why sharks and rays are so vulnerable to intense fishing pressure. Most species are near the top of the food chain and are not abundant. Sharks grow slowly, mature late in their lives, do not reproduce every year, and have few young. They succeed in part because they are long-lived. As Merry Camhi of the Audubon Society points out, "Unlike most bony fishes in which the survival of millions of eggs and larvae are often largely dependent on environmental variables, chondrichthians (sharks and rays) exhibit a much closer relationship between the number of young produced and the number of breeding adults." Kill a substantial proportion of adults, and the population will not be sustained.

Take, for example, the sandbar and scalloped hammerhead sharks, two species often caught in the Atlantic and Gulf of Mexico fisheries. The hammerhead is also frequently caught by fishermen in the small fishing camps on the shores of the Gulf of California. The sandbar takes up to sixteen years to reach maturity and then gives birth to only eight to thirteen pups every other year. Some female hammerheads can take fifteen years to reach maturity and then give birth to only twelve to forty pups per litter every other year. The Atlantic cod, by contrast, reaches reproductive maturity in only two to four years, produces 2-to-11 million eggs, and reproduces every year. The hammerhead can live to an age of thirty-five years, the cod twenty-plus years.

The impact of catching even a few of the really large predatory species of sharks—the bull, tiger, and white shark—can be even greater. For instance, when four large white sharks were caught on October 5, 1982, close to shore at the South Farallon Islands near San Francisco, the number of attacks observed on prey animals in the surrounding waters dropped by half during the next two years.

Why would the white shark be so vulnerable to fishing pressure? First, it is an apex predator, occupying the pinnacle of the food chain. Individuals of these species are never very common. The white shark feeds upon seals and sea lions, which in turn feed on smaller prey species such as fish and squid; these latter feed upon even smaller planktonic animals, and these animals feed upon planktonic plants. At each of the links in this food chain, some energy is lost, resulting in less animal mass (biomass) at each higher trophic level.

There are just not many white sharks in any one place. Roughly only a dozen identified white sharks, and probably an equivalent number that went unidentified or unseen, visited the South Farallon Islands, the best pinniped feeding grounds on the West Coast, annually during the five-year observation period from 1988–1992. The same white sharks returned over and over again at the same time. For instance, Cut Caudal, easily recognized from the missing upper lobe of its caudal fin, was observed feeding on pinnipeds off Southeast Farallon Island in four of five years. It was first seen feeding in the center of Mirounga Bay on October 23, 1988. It was then observed feeding off the northern and southern points on West End on October 10, 1989, and October 8, 1990, and two years later on October 7, 1992. Cut Dorsal, recognized by its missing upper dorsal fin, was observed at the island on all five successive years, and was lured to the surface by a surfboard on three successive years.

The true size of the local white shark population has been estimated in only two geographical areas, South Africa and South Australia. Scientists achieved these estimates by marking sharks and noting the proportion of marked to unmarked sharks when returning to the marking sites. The sharks were tagged after being attracted to boats by chumming. The number of sharks along 1,700 kilometers of coastline of South Africa from Richards Bay to Struis Bay was estimated to be 1,279 sharks (within a certain range of estimation). Another 40 white sharks were marked at Dangerous Reef and the

Neptune Islands during several expeditions sponsored by the Cousteau Society to Spencer Gulf in South Australia. This team of scientists estimated that 192 sharks existed within a 260-square-kilometer area during the second expedition and 18 sharks during the third expedition a year later, raising a concern as to whether sport fishing for white sharks was reducing their population size. Their small number and the predictability of their movements make them vulnerable to fishing pressure.

There are other reasons why white sharks are so rare. First, individuals of the species grow slowly. Male white sharks become sexually mature at a length of 11 feet 6 inches to 13 feet 6 inches and at an age of nine to ten years, when they first develop rigid claspers. A male of this size has nine to ten concentric growth rings on each cartilaginous element of the vertebral column, indicating an age of nine to ten years. Females mature at a larger size, between 14 feet 9 inches and 16 feet 5 inches, and an average age of twelve to fourteen years. It is probable that the white shark gives birth every two or three years like most other sharks. Finally, the maximum size of a litter of pups is only ten per female. In the context of this life history, it is not surprising that the capture of four white sharks at a single site would reduce the local population to half its former level. The fear of the negative impact that more thrill-seeking fishermen might have on the white shark population off California led during the first week of October 1993 to the passage of a bill in the Californian legislature protecting white sharks.

By the end of the 1980s, scientists began to voice concern over the unmanaged expansion of the shark fisheries in the Atlantic Ocean, considering the vulnerability of the species to overfishing. In 1989, the National Marine Fisheries Service began to develop a management plan for the sharks of the Atlantic Ocean and the Gulf of Mexico. This plan was implemented in 1993 as the Fishery Management Plan for Sharks of the Atlantic Ocean. The plan was directed at the management of thirty-nine species, which were divided into three categories: large coastal sharks, small coastal sharks, and pelagic sharks.

The spiny dogfish, although captured in great numbers, was not included in this plan. This is unfortunate, for the species has the same life history properties as many of the species listed. However, state Fishery Management Councils have recently put together a management plan for this species. Quotas have been set for the commercial fisheries and bag limits for the recreational fisheries for the large coastal and pelagic sharks to lessen the fishing pressure on these species. Commercial fishermen are required to hold a federal permit to fish for sharks. Significantly, fishermen are required to report the number of each species captured each fishing trip. This latter regulation will permit the National Marine Fisheries Service to monitor the catch per species on an annual basis and regulate fishing pressure based upon knowledge of whether each species catch is increasing or decreasing annually. Finally, the plan prohibits the wasteful practice of "finning," by which only the fins are retained and the rest of the body is discarded as by-catch.

The state of our knowledge about sharks must improve if the growth of shark fisheries is to continue rather than collapse. First of all, adequate identification guides are needed so that accurate fishery statistics can be collected. Then accurate records must be kept of the catch and effort of the fishermen in the fisheries. I am very excited that my graduate student, John Richert, and my Mexican colleagues are getting persons in fishing camps and sport-fishing fleets to keep such records and turn them over to scientists to keep in a database for the Gulf of California. To succeed, fishery management must be international in nature, as many of the species are highly migratory and travel across jurisdictional boundaries, making the collection of standardized fishery statistics difficult. Finally, the long life span and slow growth of these species make it impossible to assess the effect of management strategies until they have been in place for decades. Many management tools are available and currently being used in different countries, yet comprehensive management plans for sharks and rays are in place in only a few countries. These tools include the establishment of quotas, restriction of entry into the fisheries by issuing licenses, closures of geographical areas used as shark nurseries, fishing seasons, shark size and gear restrictions, and bag limits. Hopefully, in the future our important shark fisheries will be managed properly so that the mistakes of the past are not repeated.

Many of us would like the population of hammerhead sharks in the Gulf of California to be restored to its previous level. If this is not possible, it is important to return the population to a sustainable level—remaining constant from year to year. The first symptom of a collapsing fishery is the rarity of older fish in the catches of fishermen—the fishing pressure has become so intense and effective that few fish escape early capture and survive to an old age. Faced with the scarcity of adults, fishing camps along the shores of the Gulf of California have begun catching large numbers of juveniles in their gill nets. This practice will simply worsen the plight of the species, and soon there will be few juvenile hammerhead sharks as well in the gulf. As the fishing camps profit marginally from the capture of a few juvenile hammerheads, the recreational-diving industry operating out of La Paz, the largest city in southern Baja, has been impacted greatly by the rarity of adult hammerheads in the local waters. Recreational scuba- and free-divers come from many countries around the world to marvel at the famous schools of hammerhead sharks. Local dive boats take these ecotourists out to the Espíritu Santo Seamount to view hammerheads during the summer and early fall. These visitors now often leave La Paz disappointed because they saw few hammerheads.

What steps can be taken to protect the hammerhead shark? Fisheries have been managed traditionally by increasing or decreasing fishing effort based on the size of the latest catch—effort is curtailed when the catch is smaller. However, we are now confronted with declining fish stocks in all of the oceans despite vigorous attempts to manage them in this traditional manner. A new approach toward sustaining the populations of oceanic fishes is the creation of marine reserves, where fishing is completely prohibited.

The recent decline of the abundance of fishes has led U.S. and Mexican scientists to recommend the establishment of marine reserves in the Gulf of California. However, it is imperative to form these reserves based on sound scientific knowledge of the life histories of the resident species. Studies need to be directed at answering the following questions. Where should these reserves be situated? How large should they be? How many reserves are needed to protect species that migrate over great distances? A habitat of obvious importance in the gulf is the offshore water surrounding seamounts and islands, which serve as aggregating areas for a large assemblage of oceanic fishes. These locales are home to many charismatic species, such as the sailfish, several species of marlin and tuna, manta rays, whale sharks, several species of reef shark, and the scalloped hammerhead. A prudent first step in protection would be to define the north-south migratory pathway of the scalloped hammerhead shark in the eastern Pacific Ocean. This species can be viewed as a marker (a tag, in a sense), which is indicative of the presence of the entire assemblage the large assemblage of fishes occurring at the seamount. Describe the hammerhead shark's movements, and you will likely describe the migratory pathway of the accompanying assemblage of fishes.

I recently received a research grant from the Committee for Research and Exploration of the National Geographic Society to study the migratory behavior of hammerheads relative to the establishment of marine reserves. The objective of this study will be to compare the long-distance tracks of hammerhead sharks to seafloor topography to ascertain whether this important assemblage of fishes migrates between subtropical and tropical zones by moving from island to seamount, using them as stepping-stones. If the sharks are shown to move from seamount to seamount, resource managers will have a reason to prohibit fishing in small conservation zones around seamounts and islands.

Two of my graduate students, Salvador Jorgensen and John Richert, and I will make research expeditions this year to the Gulf of California. The summer trip will be devoted to determining the feasibility of using pop-up archival satellite tags on adult hammerhead sharks. These sophisticated electronic tags stay on the animal for up to a year, release from it, float to the surface, and transmit their information to a satellite. We will make free-dives into the schools of adult hammerheads, as I have done in the past at El Bajo Espíritu Santo, and tag them with this type of tag. Each tag will provide a series of geographic positions of an adult over a period of a year together with records of swimming depths and surrounding water temperatures.

Let me now place the risk of shark attacks on humans in the context of the danger human fishing imposes on shark populations. Sharks attack humans for more than one reason. Hunger is the first reason that comes to mind. In a hunger-motivated attack, one would expect that the shark would consume most of its victim. Indeed, this is the case in some attacks on humans. In October 1939, two divers were observed being attacked near a beach of New South Wales, Australia, and during the following day their remains were taken from an 11-foot-4-inch tiger shark.

Alternatively, the wounds inflicted in attacks by other, smaller species of shark often consist of only crescent-shaped bite marks, little flesh being removed from the victim. This observation led David Baldridge and Joy Williams to write an influential article in 1969 entitled "Shark Attack: Feeding or Fighting?" In this article, they argued that "attacking sharks were motivated by pressures other than hunger such as territorial behavior or unintentional interference by the victim in courtship." They reported that in one case the shark had been photographed before the attack, and the shark's manner of swimming was very erratic and unusual. The pectoral fins were extended more downward than usual, with the nose pointed upward, and the back hunched. The shark appeared to swim stiffly with its whole body, the head moving back and forth almost as much as the tail.

Richard Nelson, a graduate student at the California State University at Long Beach and Don Nelson later photographed and analyzed this aggressive display in the gray reef shark, a species common in the South Pacific. Don would charge toward a gray shark holding a powerhead in one hand for protection while Richard filmed the shark's response to his charge. They broke down the display into two locomotory elements and four postural elements for the sake of understanding it better. The locomotory components were exaggerated swimming movements to the side on a horizontal plane and swimming in a looping configuration. The four postural components were an upward pointing of the snout, a lowering of the pectoral fins, the arching of the back, and the bending of the tail in a lateral direction. I have observed this same display in the blacknose, blacktip, silky, and bonnethead sharks, and would be surprised if the display were not exhibited in some form by every shark species. The intensity of the display (and likeliness of attack) is proportional to the degree to which the diver confines the shark and the speed at which that person approaches. For this reason, the display can be avoided if a diver swims slowly away from the displaying shark.

Baldridge and Williams were also aware of the grab-and-spit behavior of the white shark. They mentioned the case of a skin-diver in Bodega Bay, California, who "was recently struck without warning by a very large shark. The victim stated that he felt 'something grab his leg' and clamp down 'like a giant vise,' crushing his back and chest." He also said, "I could see it was a shark, so I just went limp and played dead, and finally it let go." The victim was carried about 10 feet and released without the shark showing any further interest in him. They concluded that the action was not consistent with efforts to feed on the man. I would suggest that the shark may have rejected the victim, because he was too low in fat.

The risk of being injured by a shark attack is low relative to other sources of injury in our everyday environment. The number of attacks worldwide by all species of sharks averaged 46.7 per year during the decade of the 1990s; the attacks by white sharks averaged 7.2 per year. The majority of these attacks were not fatal. Information about them is kept in the International Shark Attack file. This archive resides at the Florida Museum of Natural History of the University of Florida, where it operates under the auspices of the American Elasmobranch Society. New observations and historical records are constantly added to this database by a worldwide network of cooperating investigators.

The number of unprovoked attacks by sharks on humans has increased over the years. For example, the number of unprovoked white shark attacks per decade rose from four attacks during the ten-year period from 1900 to 1909 to seventy-two attacks from 1990 to 1999. However, the increase in the frequency of attacks is likely not due solely to growing white shark populations, but also to underreporting during the early years and greater recreational use of the coast in recent years.

The chance of being hurt by a shark is insignificant relative to far greater hazards in our society. During the last year of the 1990s, there were over 40,000 deaths from automobile crashes in the United States, 14,000 fatalities from accidental falls, 791 drownings in pools or at beaches, 296 drownings in bathtubs, and 72 deaths by lightning. Other animals—one considered our best friend—are the cause of more human fatalities than sharks. For example, in the United States there are 220 deaths per year from accidents related to riding horses, 50 from the stings of hornets, wasps, or bees, and 10 from dog bites. The fewer than 50 shark attacks occurring around the world each year seem less significant in the context of these other common risks. In actuality, you expose yourself to greater danger when you drive to the beach than when you swim in the waters off the beach. It is only by learning more about the behavior of sharks by watching them in aquariums or viewing them in their own environment that we will overcome our fear of them.

SOURCE MATERIALS

When writing this book, I often based my narration on daily entries in my laboratory and field notebooks, completed at a rate of roughly one per year over the last twenty-nine years of my professional life. Other sections of the book are based on the contents of articles, published in the popular and scientific literatures. If you interested in learning more about a particular study described in this book, contact the librarian at your nearest university library, which has a collection of marine journals. That person will tell you how to obtain a copy of the article. The references for these articles are given below, under each chapter of the book.

1. Shark Fever
Klimley, A. P. 1974. An inquiry into the causes of shark attacks. *Sea Frontiers* 20:66–75.

2. Cross-Species Dressing
Cummings, W. C., and P. O. Thompson. 1971. Gray whales, *Eschrichtius robustus*, avoid the underwater sounds of killer whales, *Orcinus orca*. *Fishery Bulletin* 69:525–30.

Klimley, A. P., and A. A. Myrberg, Jr. 1979. Acoustic stimuli underlying withdrawal from a sound source by adult lemon sharks, *Negaprion brevirostris* (Poey). *Bulletin of Marine Science* 29:447–58.

Myrberg, A. A., Jr., C. R. Gordon, and A. P. Klimley. 1978. Rapid withdrawal from a sound source by open ocean sharks. *Journal of the Acoustical Society of America* 64:1289–97.

———. 1976. Attraction of free-ranging sharks by low-frequency sound, with comments on its biological significance. Pp. 205–39 in A. Schuijf and A. D. Hawkins, eds., *Sound reception in fishes*. New York: Elsevier Press.

Nelson, D. R., and S. H. Gruber. 1963. Sharks: Attraction by low-frequency sounds. *Science* 142:975–77.

3. Shark Sex in the Miami Seaquarium

Clark, E. 1959. Instrumental conditioning of lemon sharks. *Science* 130:217–18.

Klimley, A. P. 1978. Nurses at home and school. *Marine Aquarist* 8:5–13.

———. 1980. Observations of courtship and copulation in the nurse shark, *Ginglymostoma cirratum*. *Copeia* 1980, 878–82.

Klimley, A. P., and A. A. Myrberg, Jr. 1979. Acoustic stimuli underlying withdrawal from a sound source by adult lemon sharks, *Negaprion brevirostris* (Poey). *Bulletin of Marine Science* 29:447–58.

Pratt, H. L., Jr., and J. C. Carrier. 2001. A review of elasmobranch reproductive behavior with a case study on the nurse shark, *Ginglymostoma cirratum*. *Environmental Biology of Fishes* 60:157–88.

4. Diving with Sharks in Southern California

Tricas, T. C. 1979. Relationships of the blue shark, *Prionace glauca*, and its prey species near Santa Catalina Island, California. *Fishery Bulletin* 77:175–82.

5. Swimming with Hammerhead Sharks

Klimley, A. P. 1981. Grouping behavior in the scalloped hammerhead. *Oceanus* 24:65–71.

———. 1995. Hammerhead city. *Natural History* 104:32–39.

Klimley, A. P., and D. R. Nelson. 1981. Schooling of scalloped hammerhead, *Sphyrna lewini*, in the Gulf of California. *Fishery Bulletin* 79:356–60.

———. 1984. Diel movement patterns of the scalloped hammerhead shark (*Sphyrna lewini*) in relation to El Bajo Espíritu Santo: A refuging central-position social system. *Behavioral Ecology and Sociobiology* 15:45–54.

6. Solving the Mystery of Hammerhead Schools

Klimley, A. P. 1985. Schooling in the large predator, *Sphyrna lewini*, a species with low risk of predation: a non-egalitarian state. *Zeitschrift fur Tierpsychology (=Ethology)* 70:297–319.

———. 1987. The determinants of sexual segregation in the scalloped hammerhead, *Sphyrna lewini*. *Environmental Biology of Fishes* 18:27–40.

Klimley, A. P., and S. T. Brown. 1983. Stereophotography for the field biologist: Measurement of lengths and three-dimensional positions of free-swimming sharks. *Marine Biology* 74:175–85.

7. Shark Rush Hour at Gorda Seamount

Cigas, J., and A. P. Klimley. 1987. A microcomputer interface for decoding telemetry data and displaying them numerically and graphically in real time. *Behavioral Research Methods, Instruments, and Computers* 19:19–25.

Klimley, A. P. 1993. Highly directional swimming by scalloped hammerhead sharks, *Sphyrna lewini*, and subsurface irradiance, temperature, bathymetry, and geomagnetic field. *Marine Biology* 117:1–22.

Klimley, A. P., I. Cabrera-Mancilla, and J. L. Castillo-Geniz. 1993. Descripción de los movimientos horizontales y verticales del tiburón martillo *Sphyrna lewini*, del sur de Golfo de California, México. *Ciencias Marinas* 19:95–115.

8. Hammerhead Sharks as Ocean Navigators

Galvan-Magaña F., H. Nienhuis, and A. P. Klimley. 1989. Seasonal abundance and feeding habits of sharks of the lower Gulf of California. *California Fish and Game* 75:74–84.

Klimley, A. P., and S. B. Butler. 1988. Immigration and emigration of a pelagic fish assemblage to seamounts in the Gulf of California related to water mass movements using satellite imagery. *Marine Ecology Progress Series* 49:11–20.

Klimley, A. P., S. B. Butler, D. R. Nelson, and A. T. Stull. 1988. Diel movements of scalloped hammerhead sharks (*Sphyrna lewini* Griffith and Smith) to and from a seamount in the Gulf of California. *Journal of Fish Biology* 33:751–61.

9. In Quest of the White Shark

Ainley, D. G., C. S. Strong, H. R. Huber, T. J. Lewis, and S. H. Morrell. 1981. Predation by sharks on pinnipeds at the Farallon Islands. *Fishery Bulletin* 78:941–45.

Klimley, A. P. 1987. Field studies of the white shark, *Carcharodon carcharias*, in the Gulf of the Farallones National Marine Sanctuary. Pp. 33–36 in M. M. Croom, ed., *Current Research Topics in the Marine Environment*. San Francisco: Gulf of the Farallones National Marine Sanctuary.

Klimley, A. P., and S. D. Anderson. 1996. Residency patterns of white sharks at the South Farallon Islands, California. Pp. 365–73 in A. P. Klimley and D. G. Ainley eds., *Great White Sharks: The Biology of* Carcharodon carcharias. San Diego: Academic Press. 528 pp.

Mollet, H., G. M. Cailliet, A. P. Klimley, D. A. Ebert, A. T. Testi, and L. J. V. Compagno. 1996. A review of length validation methods for large white sharks. Pp. 91–108 in A. P. Klimley and D. G. Ainley, eds., *Great White Sharks: The Biology of* Carcharodon carcharias. San Diego: Academic Press. 528 pp.

Randall, J. E. 1973. The size of the great white shark. *Science* 181:169–70.

10. White Shark Predation at the Farallon Islands

Anderson, S. D., A. P. Klimley, P. Pyle, and R. H. Henderson. 1996. Tidal height and white shark predation at the South Farallon Islands. Pp. 275–79 in A. P. Klimley and D. G. Ainley, eds., *Great White Sharks: The Biology of* Carcharodon carcharias. San Diego: Academic Press. 528 pp.

Klimley, A. P. 1994. Do white sharks (*Carcharodon carcharias*) select prey based on high-fat content? Transcript of scientific talk, Meeting of the American Association of Ichthyologists and Herpetologists (ASIH), University of Southern California.

Klimley, A. P. 1994. The predatory behavior of the white shark. *American Scientist* 82:122–33.

Klimley, A. P., S. D. Anderson, P. Pyle, and R. P. Henderson. 1992. Spatiotemporal patterns of white shark (*Carcharodon carcharias*) predation at the South Farallon Islands, California. *Copeia*, 1992, 680–90.

Klimley, A. P., P. Pyle, and S. D. Anderson. 1996. The behavior of white shark and prey during predatory attacks. Pp. 175–91 in A. P. Klimley and D. G. Ainley, eds., *Great White Sharks: The Biology of* Carcharodon carcharias. San Diego: Academic Press. 528 pp.

11. Talking with Their Tails

Bruce, B. D., G. H. Burgess, G. M. Cailliet, O. B. Gadig, and W. Strong. 1994. The evolutionary significance of upper caudal lobe Tail Flap behaviour in the white shark (Lamnidae, *Carcharodon carcharias*): Alternative hypotheses. Unpublished manuscript.

Klimley, A. P., P. Pyle, and S. D. Anderson. 1993. Displays and intraspecific competition among white sharks (*Carcharodon carcharias*). Transcript of scientific talk, Symposium on the Biology of the White Shark, *Carcharodon carcharias*, Bodega Bay, California.

———. 1996. Is the Tail Slap an agonistic display among white sharks? Pp. 241–55 in A. P. Klimley and D. G. Ainley, eds., *Great White Sharks: The Biology of* Carcharodon carcharias. San Diego: Academic Press. 528 pp.

12. Baby White Shark Gets Away on National Television

Compagno, L. J. V. 1984. FAO Species Catalogue. Vol. 4. *Sharks of the World: An Annotated and Illustrated Catalogue of Shark Species Known to Date.* Part 1. Hexanchiformes to Lamniformes. *FAO Fisheries Synopses* 125 (vol. 4, pt. 1):1–249.

———. 1984. FAO Species Catalogue. Vol. 4. *Sharks of the World. An Annotated and Illustrated Catalogue of Shark Species Known to Date.* Part 2. Carcharhiniformes. *FAO Fisheries Synopses* 125 (vol. 4, pt. 2):251–655.

Klimley, A. P. 1985. The areal distribution and autoecology of the white shark, *Carcharodon carcharias*, off the western coast of North America. *Southern California Academy of Sciences, Memoirs* 9:15–40.

Klimley, A. P., S. C. Beavers, T. H. Curtis, and S. J. Jorgensen. 2002. Movements and swimming behavior of three species of sharks in La Jolla Canyon, California. *Environmental Biology of Fishes* 63:117–35.

Reidarson, T. H., and J. McBain. 1994. Hyperglycemia in two great white sharks. International Association for Aquatic Animal Medicine, *Newsletter* 25 (October issue):6–7.

13. Electronic Monitoring of White Sharks at Año Nuevo Island

Boustany, A. M., S. F. Davis, P. Pyle, S. D. Anderson, B. J. Le Boeuf, and B. A. Block. 2002. Expanded niche for white sharks. *Nature* 415:35–36.

Klimley, A. P., F. Voegeli, S. C. Beavers, and B. J. Le Boeuf. 1998. Automated listening stations for tagged marine fishes. *Marine Technology Journal* 32:94–101.

Klimley, A. P., B. J. Le Boeuf, K. M. Cantara, J. E. Richert, S. F. Davis, and S. Van Sommeran. 2001. Radioacoustic positioning: A tool for studying site-specific behavior of the white shark and large marine vertebrates. *Marine Biology* 138:429–46.

Klimley, A. P., B. J. Le Boeuf, K. M. Cantara, J. E. Richert, S. F. Davis, S. Van Sommeran, and J. T. Kelly. 2001. The hunting strategy of white sharks at a pinniped colony. *Marine Biology* 13:617–36.

14. Shark-Eating Humans or Man-Eating Sharks?

Baldridge, H. D. J., and J. Williams (1969). Shark attack: Feeding or fighting? *Military Medicine* 134:130–33.

Burgess, G. H. 1999. Statistics on shark attacks in the United States and worldwide. International Shark Attack File, Florida Museum of Natural History, University of Florida, Gainesville (personal communication).

Camhi, M. 1998. *Sharks on the Line: A State-by-State Analysis of Sharks and Their Fisheries.* New York: Audubon Society.

Camhi, M., S. Fowler, J. Musick, A. Bräutigam, and S. Fordham. 1998. Sharks and their relatives: Ecology and conservation. Occasional paper. *IUCN Species Survival Commission* 20:1–39.

Johnson, R. H., and D. R. Nelson. 1973. Agonistic display in the gray reef shark, *Carcharhinus menisorrah*, and its relationship to attacks on man. *Copeia*, 1973, 76–84.

Klimley, A. P. 1974. An inquiry into the causes of shark attacks. *Sea Frontiers* 20:66–75.

———. 1999. Sharks beware. *American Scientist* 87:488–91.

National Safety Council. 1999. Total deaths due to injury, United States, 1993–1995 (deaths averaged for three years), Query of Worldwide Web.

Nelson, J. S. 1976. *Fishes of the World.* New York: John Wiley & Sons.

INDEX

A. PETER KLIMLEY is an internationally known marine biologist. An adjunct associate professor in the Department of Wildlife, Fish, and Conservation Biology at the University of California, Davis, Dr. Klimley has published articles on sharks in *American Scientist*, *Natural History*, and other popular magazines, as well as more than fifty scientific articles. He has appeared in numerous film documentaries worldwide. A coauthor of the leading academic book on great white sharks, Dr. Klimley lives in Petaluma, California.